MARC HAMER

Marc Hamer lived in the North of England before moving to Wales. After spending a period homeless, then working on the railway, he returned to education and studied fine art in Manchester and Stoke-on-Trent. His previous book is *A Life in Nature: Or How to Catch a Mole* and was longlisted for the Wainwright Prize.

MARC HAMER

Seed to Dust

A Gardener's Story

VINTAGE

1 3 5 7 9 10 8 6 4 2

Vintage is part of the Penguin Random House group of companies
whose addresses can be found at global.penguinrandomhouse.com

Copyright © Marc Hamer 2021

Illustrations by Jonathan Ashworth / agencyrush.com

Marc Hamer has asserted his right to be identified as the
author of this Work in accordance with the Copyright,
Designs and Patents Act 1988

First published in Vintage in 2022
First published in hardback by Harvill Secker in 2021

penguin.co.uk/vintage

A CIP catalogue record for this book is
available from the British Library

ISBN 9781529112498

Printed and bound in Great Britain by Clays Ltd, Elcograf S.p.A.

The authorised representative in the EEA is Penguin Random House
Ireland, Morrison Chambers, 32 Nassau Street, Dublin D02 YH68

This book is substantially a work of non-fiction based on the life,
experiences and recollections of the author. In some limited cases,
names of people and places and the sequences of events have been
changed solely to protect the privacy of others.

Lines on p.228 from 'The Love Song of J. Alfred Prufrock' by T. S.
Eliot; lines on p.291 from 'Burnt Norton' by T. S. Eliot; line on p.384
from 'Do not go gentle into that good night' by Dylan Thomas; With
thanks to The John Clare Society for the lines on p.217 and p.355
from *Clare: selected poems and prose* by John Clare.

Penguin Random House is committed to a sustainable future
for our business, our readers and our planet. This book is made
from Forest Stewardship Council® certified paper.

This book, like my life, is for Peggy.

Contents

Prologue

The swifts have left the bell tower and are on their way to Africa.

As an exercise in holding my attention on a single thing, my mind's eye held the pattern that an individual made as a wandering pencil line across the sky. Another swift crossed the line, and then more, returning as all things do to the creative chaos from which they came. In the warp and weft of the strands they drew, I saw the structure of this book – cycles around cycles, lives lived, relationships made and lost, from seed to dust.

Written in a tradition as old as storytelling itself, in essence what is here is truth, although in fact it is often not. What follows is drawn from memory and, just like any other drawing, any other memory, perspectives are distorted, time is contracted and the sun shines in the imagination where in reality there was only shade.

Villedieu les Poêles, Northern France

January

White

Fallen leaves curl as if to fold their fingers in for warmth. Hot breath steams from the rudely open gobs of pipes on the outside walls of houses, while inside hungry, roaring flames or coiled electric elements nested deep in boilers keep the people safe, away from nature's icy teeth. The air outside is densely filled with crystals that turn it into mist so thick I cannot see the nearby church spire through the leafless trees. All is still. Still and silent. The celebrations of Christmas and New Year seem long gone; the people who have jobs have returned to them, but I have not. I'll have this month for myself, as gardeners often do. January is a time for looking at seed catalogues and dreaming of what could be, if I moved this and dug up that, and planted those over there. The garden floating in my mind shifts like a Mark Rothko painting as I slide a block of colour here and merge two colours there, make a path to break things up, plant a scribbly hedge to build a magnetic corner space. All gardeners have fantasy gardens, and many of them are painters. I no longer paint: painting requires lots of equipment and a permanent workspace to keep it all. I write instead. I can do that anywhere.

The world, outside my house and in, seems peaceful. Although the world of men is rarely peaceful, my own small bounded world relaxes. In the Rookwood where I live, everything looks black and white. The jackdaws, slow and calm, huddle by the chimneys or jab half-heartedly at the frozen earth, the lucky ones pull listless cold worms from the ground or squirming arthropods. The trees do not sway, but point their limbs up and around and wait. The sparrows are quiet, too, still flitting from shrub to shrub but not saying much, and I am watching through my window as if waiting, but not waiting.

The cold is busy doing vital work, sneaking between the grains of earth, lowering the temperature of molecules of water so they slow and cease their movement then expand and push the grains apart, so when the thaw arrives, the clods of soil on the surface crumble. It creeps into the cells of beasts and mingles with their being, as they breathe out warmer air that condenses into steam. It seeps into poor houses and chills the feet and clothes of warm children dressing for school; clings onto the home-less as they shelter in their doorways overnight; and builds crystals on the edges of tawny, dead hydrangea petals, kissing the kale and sprouts and winter cabbages to make them sweet and tasty. It wraps thickly around the apple trees to send them into a sleep so deep that, when they wake, they burst with energetic fruits.

I'm indoors and resting like my cat, who jumps onto my lap as soon as I sit down. A tortoiseshell called Mimi, who loves me like I love her, selfishly and greedily wanting my warmth, as I want her adoration and luxury. She looks into my face; she has splotches in her eyes, brown spatters against the amber, like freckles. I've always been a sucker for freckles. I'm torn between the book I'm reading and watching the outside go by. I'm reading, yet again, W. G. Sebald's *Rings of Saturn* and feeling snug and blanketed as I follow his meandering journey, which seems to me to come from nowhere and end up nowhere. I like a story that feels real like that.

Beginnings

A new year, a new calendar, a new diary. The indoor world doesn't feel so new; the same dust blows under my desk, the same ache in my left knee. The only thing that's new is my diary, leaning next to the old one, which still has a few entries I need to copy across. Had the old diary more pages, it would carry on being used and would do its job as well. It needn't end quite where it does.

In the distant past, our ancestors chose to begin the new year at midwinter, when the feast of harvest was long gone and there were no available crops in the field, while they watched for the spring to come and show some mercy. Perhaps, in more primitive times, people may have feared that winter was the end of a world that was fading away into permanent cold and darkness. Somebody wrapped warmly in skins perhaps became aware that the falling sun had changed its behaviour and was now climbing higher each day. 'Look, guys, it's going to be okay!' They watched it all go by and saw that everything in the world was always changing, all at different rates. There's always something new appearing on the scene, coming round the corner. Zoom in too close and things appear to pop into existence and then pop out again. Move back and you can see it all spins round and everything is just background whorl. When

I'm feeling sad for any reason, I've learned to take a step back, but our senses can only perceive so much; there is so much more they cannot see, and we can never know what exists outside our narrow frame.

After the longest night of the year around 22 December, when the sun slowly starts to return, it seems a good time to begin a new cycle. So we mark the rim of the wheel that infinitely turns; we make it halfway through the darkness as it fades into the light, and we say, 'This is where the circle starts.' That day was the very first date in the very first calendar and the start of our culture. I wonder what human life on this planet might have been like, had we been unafraid of the dark and felt no need to count the days and the seasons; if we had never developed a system of numbers, like some Amazonian tribes and children who do not recognise the difference between three sweets and four.

Our culture is rooted in us differentiating between things: night from day, lunchtime from breakfast, us from them, good from bad. Balancing one thing against another. We learn to isolate things from their continuously connected nature and to create clear beginnings and endings. The ancient circular symbol of a serpent eating its tail, the ouroboros, has the head and tail in the same place, but clearly has a beginning and an end, although it goes on for ever. Zen Buddhists have a similar image in the *ensō*, a simple circle, small or massive; painted in one or two strokes by a master of calligraphy, it starts with the brush being placed somewhere on the page and goes round, then ends at the same place. It symbolises the cycle of life, often wobbly and imperfect – there is nothing outside the circle, there is nothing inside. It is said that the character of the creator can be inferred from the way he paints this circle.

Peppered Moth

In May and June a common peppered moth lays around 2,000 tiny, soft white eggs the size of a pin-head, hidden in deep cracks, high in the canopy, in the bark of a tree. A couple of weeks later each surviving egg splits as the contents grow, and a caterpillar crawls out. It eats its own leathery egg, then the soft parts of the nearest leaf. It eats until it is a thousand times bigger. Birds and bats forage for it high in the leaves and branches. This is an 'instar', a stage in the development of a small creature before it sheds its final skin and becomes an adult; caterpillars and nymphs and tadpoles are 'instars'. This peppered moth instar was one of the first wild creatures to be identified by scientists as being naturally camouflaged, for it looks like a twig of the tree it was born on: greeny-brown, stiff and straight, at rest. The scales on the wings of the moth that it becomes are coloured to disappear against the lichen that grows on the trees where it lives, breeds, lays and dies. All the parts of its life are matched to that tree.

The instar eats and grows fat and tight, its skin splits open and a new caterpillar's arched back bursts out, pulls its head free and struggles out of the old skin, leaving it clinging onto the branch with its hollow bud-like legs. Just another bit of twig. The caterpillar sheds its skin two or three times in its life. Later in the year, around October, as fat as it can get, it quits the leaves where it was born, works its bumbling way down to the earth and burrows into the soil, where it lies over the winter, pupating. It creates a hard brown shell, and inside that little house the body of the caterpillar breaks down into a disorganised brown mush that squirms unpleasantly if you touch it. Cell soup in a natural test tube, which churns and reinvents itself before the spring comes.

It grows in size and in April or May it wriggles to the surface, where the dry cocoon splits open and a new creature bursts out from its weird sleep. With long and jointed legs, it hangs onto a leaf or twig. No longer a crawling, burrowing thing, it pumps up its wings with blood and they dry in the sun and wind and it flies up. The female flies only once in her lifetime; she waits high in the tree she fell from in a previous life, and sends out pheromones to attract a male. The males will fly every night until they find a mate and will rest in the trees during the day. After mating, the males will stay with her to protect her from other male moths until she lays her many eggs, deep in the cracks in the bark, while the birds and bats look to take them. By the end of the summer the moth will have died. It has never made a decision in its life; it just does what it does because that is what it does. It follows the flow that takes it where the flow always goes.

I start my day as it breaks and I write about it often. Drawing the imperfect circle over and over again. Yesterday is the distant past and I start the day from where I am, not where I was; and here I am at breakfast writing notes, with the day ahead of me and the night behind. I think daybreak is the finest time, when she and I are warm together, cosy waking to the first unedited line of the circular poem of this day and all our days. I focus in, and every waking is a birthday, joyous, newly born. I make her tea, we chat.

Peggy, sitting by me, says, steaming cup in hand, 'Your beard is now completely white.'

I say, 'We made it through, didn't we?'

'It is so white it sparkles,' she says.

It was not always so easy.

February

Returning

Miss Cashmere is looking older. Slightly more papery. Like something Japanese and delicate. A lantern, bright, pretty. Occupying space, but so fragile she could be blown all the way back to Japan in a slight wind, turned into pulp by a small shower, dashed to the ground by the curiosity of a bird. A blue tit. A moth fluttering at her brightness. A red silk peony holds her hair up in a white bun. Straggles have escaped and those at the back curl down and then around themselves against the freckles on her neck.

It is February. I haven't seen her since the beginning of December. She has passed a winter living alone. Like a chrysalis from under the soil, she has emerged from the darkness changed: she is more bent, she is twisted like a rusty woodscrew. She is nearly eighty.

I've been in the fields all winter, catching moles and waiting for the spring, and trying to stay warm by keeping moving and pretending not to be ancient, but I am. I feel as worn as a creaky gate; I long to fall open on my hinges to the easy place where the wood hits the mud and the world passes by. Do Miss Cashmere's hinges and muscles ache? Is her breathing shallow, her heart irregular? In the empty darkness can she hear the squeals and howls of tinnitus? Can she cut her own toenails or does somebody in a white coat, a name badge on a lanyard, do it for her, kneeling on a cushion at her feet, wearing surgical gloves?

This is my first day back at work in the garden after a long winter break. There are molehills in the lawn; a white crust of frost tops the fresh soil they dug last night, little snowy mountain landscapes. When they thaw I'll rake out the hills as a top dressing on the lawn. I will no longer set traps for the moles, throw their soft bodies to the crows. I'm done with the demeaning business of killing things, a business that worked away at me little by little until I felt closed inside.

She is sitting at her kitchen table, reading the newspaper and smoking a cigarette, a flash of white hair through the window. The floating trail from her cigarette and her hair and the window seem to be made of the same ghostly fabric, here and not here. The red peony glowing mistily, the only real colour. Through the softness of merging clouds of smoke and hair, I see the weak pink skin of her scalp as I drift by the house, like fog through the curled pale flowers of wintersweet (*Chimonanthus fragrans*), yellow and brown. Some have already fallen onto the grass. We together are all just passing fumes.

These tender flowers at the edge of brown decay are the most sweet and fill me with a love that has no desire – no desire that I can understand or name, at least. To me those flowers are always falling. I've seen the little bobbles of pale buds, tight and hard like pebbles staggered along the length of their thin, bare, knobbly twigs in December, but never seen the opened buds, the soft bloom. Not these buds, not these flowers. They open in January, and I don't go into the garden in January. Nobody goes there at all. Neither she nor I saw them open and flower and bloom, and I alone saw them die. They are not there for us. The wintersweet is said to have the most lovely fragrance of any plant, but I have never smelled its fresh perfume, only its sweet decay. In full bloom it is pale, but now it's brown and twists from the branch and falls. Dried and withered. Empty seashells. Dead insects. Carapace. Crust. Wings. Cocoon. Frail and tender crisping paper, becoming soggy with rain and then slime and then earth.

The books say that the wintersweet has nothing to offer after the flowers have gone, but its reddish bark and spindly knotted stems before the leaves arrive look full of promise to me. When the green comes, the little shrubby tree, plain and pale, fades into the background until next winter when, while there is nobody watching, it will dance again, alone, to its own tune. It doesn't try to compete with the big, noisy blooms of summer, but in the winter it fills the garden with its own sweet delicate scent. When it's done, I'll cut the oldest branches to the ground with a saw and shears, so that fresh young stems can come through and flower again.

Everything is fragile in the cold of winter, ready to crack or fall and rot, but the seasons change and we roll slowly closer to the sun, like an opening eye that looks towards the warmth and light of that vital star. Living things begin to eat and fatten, to consume, so that they have the energy to make copies of themselves and grow plump to survive yet another drift away into the inevitable dark. Everything changes, and only the changing stays the same.

Ice

I feel heavy and lumpen. The smell of last year's leaves, wet and flat, floats up from the earth. Fallen from the sycamore, leaves the colour and texture of my boots layer on my laces like the thinnest leather patches as I walk through the slick grass – a foliate beast with feet of leaves under thick brown corduroy trousers that are held up with braces and turned up at the bottom. The damp old canvas bag, banging heavy at my hip, weights me further. In it a trowel, a hand-fork, a folding pruning knife with a curved blade, a sharpening stone, a coil of green garden wire, a ball of string. Bits of cracked wood, bits of rusty metal, bits of tarred and twisted hemp made into the shape of tools and worn to fit my hand.

The cold air comes in and steam leaves through my nostrils, making warm tendrils and moist clouds that I can feel on the skin of my face and that dampen my beard. Limping a little. My left knee is stiff. The winter feast and rest have made me fat, but hard work will soon get me fit again. Despite the cold and the pain, I'm happy to be back, walking on crystal-frosted grass. My feet make crushing tissue-paper sounds: crush, crush, crush.

She waves at me as through the window where she sits reading her paper, smoking. Drinking from an ugly brown mug, flicking her ash into a saucer. We haven't spoken yet this year. It wasn't a 'come and chat' kind of wave; it was a 'hello, nice to see you, don't come and chat' kind of wave. I smile and wave back as I walk on past the house, which sits above the vast garden on a stone platform, like a birdcage on a table. On my rounds. My first job, at eight-thirty, is to look the garden over to see what needs doing today. Her old tortoiseshell cat runs across the grass on the trail of something and into the undergrowth of a hedge. I'm happy to see the old cat again, starting to warm to the day and its adventures.

A low wall surrounds the house that is just high enough to sit on. A paved stone area with folding metal tables and chairs. A climbing hydrangea (*Hydrangea petiolaris*) grows on the south wall of the house and needs pruning. It still has its dead brown flower heads from last summer. Rusty and soggy.

Icicles drip off the edge of the roof, a slow pat-pat-patting onto the stone patio. The old blue-black slate roof tiles are shining as the sun reflects itself; looking into its brightness bouncing off the window makes me screw up my eyes. Magpies chatter on the ridge tiles. Jackdaws squabble on the chimneypots and I stop to look and take in a big deep breath, fill myself with the new day – cool and moist and stuffed with the scents of green, of cold air, the warmth of my body and jasmine – and I feel wonderful.

Jasmine

From the stone-flagged patio, surrounded by pierced sandstone walls, steps lead down to the sloped and terraced lawns with their flower beds towards the walled pond, where three green dolphins nose towards a foliate fountain, which, when it's on, shoots a high umbrella of rain that showers any passers-by when the wind gusts. Icicles hang from the dolphins and there is a frozen glossy sheen on the verdigris of the bronze, sitting in a mass of water that is a still and solid grey-green. On the left of the lawns is an orchard with a few apple and pear trees and benches scattered around, hidden behind tall yew hedges.

Beyond the pond, behind a beech hedge, the garden is let off the lead like a dog to run wild. The summerhouse faces the vast meadow, which is flat and bare and frosted; a few sturdy thistles, dead and white, still stand and cast long shadows, and there are molehills and tussocks of sedge where the meadow dips, and a small stream trickles from an icy spring and freezes in a swampy patch. Behind the meadow a woodland of bare trees makes the garden look as if it goes into the distant mountains. To the right of the summerhouse and tucked away behind another hedge are the stables and the compost heaps, three broken greenhouses and a gravel track where I park my van. There is no other dwelling to be seen, only far-off windmills standing still in the flat air on ridges just before the air turns blue. Around at the front of the house is a smaller lawn surrounded by flower beds planted cottage-style, and an old stone wall where roses climb, which separates the garden from the quiet road. There's a bigger house down to the right – they have horses in the field, and stables. There used to be horses here, too, but the stables are empty now; one stall houses tools, another the mowers. The garden is twelve acres. This is my daylight world. I have never been inside the house.

When the day's work is over, I close the gates, thread the heavy grey chain through the rusted, once-black painted bars and snap the padlock shut. I turn my back and leave this private world each day for another universe, my shadow-world where weekends, evenings and nights are lived with Peggy, who looks out of the window and writes stories she makes up in her head; home to my shelves filled with the thoughts of poets and thinkers. My life is simple: there is light and shade, the light is beautiful, the shade more so.

The tiny six-pointed yellow flowers of winter jasmine (*Jasminium nudiflorum*) are tangling happily on the pierced wall. Snowflakes, stars, distant faintly scented suns, a scattered galaxy anchored in space by the gravity of galvanised wires threaded through iron spikes, which I had hammered into the crumbling mortar after the weight of the plant pulled down the rotten painted wooden trellis. Red and crusty brown, oxidised iron crystals sparkle through the dull grey zinc plating of the wire, where it has cracked or worn away from the constant friction of the bright plant swaying in the breeze as it tries to escape and fall to the ground. The structure will last perhaps another four or five years, before it gives up its metal to the air and rusts so thin it breaks and needs replacing. I'll lay the plant gently on the grass in the spring after pruning, as I did a couple of years ago, take out the old wires, spin a shiny new web and lift it up, stem by stem, arranging them so they do not cross and rub at the thin, waxy single layer of cells that protects the plant from disease and decay – the epidermis. I'll tie the universe of stars to the net with green garden twine, which has that dusty, dry smell I love so much. I like things that smell. The little flowers smell, a faint scent from the hundreds of flowers that drip from their whippy green stems. A slow firework, a universe that takes a year to explode and seems over in a moment. What does the universe smell of? Mine smells of green and oil, old books, cold air, warm body. Here I'm happy, my days are lived in contentment.

Whatever lives in the earth – whatever great beast pushes these living things out and makes the world sing – is still sleeping, but as the world warms, she will wake and things will begin to grow. To me, it is a she because that is in my nature, but of course the beast is all possible genders at once. It is only humans that define and name things; nature doesn't waste its time on that. I pass underneath a huge horse-chestnut tree and the buds are sticky and pointed, glossy brown, waiting to unwrap. The closed, protective fists of leaves will soon uncurl into a million five-fingered solar panels, open hands that spread to grab

25

the rays and bring them home to turn them into sugars, so it can grow and make its massive shiny seeds that children love.

In the soil a little flash of fresh green shows through the ice crystals that have grown on the tiny raised peaks of tilled earth. A curled fern crozier – a fiddlehead – nestles with its companions, furry and waiting: a ram's horn, a baby goat buried, sparkling ice growing on its hairy back. Yellow and blue crocus are just pushing their closed heads through a lawn that won't be mowed for weeks. How can something so papery and tender move through the hard, frozen earth without damaging itself?

As each living thing slowly eases into a new state of being, it appears as something new and then expands and becomes something else, then something else again. Sometimes visibly changing before my eyes: the daisies that open when the sun rises, then close again at dusk. Life comes in, life goes out, life comes in again. The beast breathes. The perennial plants spread and find their places, then disappear for a few months, to return again in their predictable rhythmic, cyclical way. Each time claiming a little more space.

Away from the flowers, there is no scent on the air other than the cold, no sensation on my skin other than the cold. The light is grey and low and flat and cold. There are no shadows. The trees are bare and still and cold, and the hairs on my arms are standing straight up.

Another Gardener

The kitchen door slams as I walk away to get the ladder and Miss Cashmere drives her big old green Jaguar down the drive. She has had that car for years. I remember when it was new. Together we live different lives in this place. I arrive and leave unseen, through massive old wrought-iron gates set in a wall of trees half a mile from her house, and drive down a narrow track to the greenhouses and compost heaps. You've seen gates like that at the side of the road, you've wondered what's behind them. They really are the entrance to the wonders you imagined. She comes and goes through a varnished five-barred gate at the top of a gravel drive that leads to the front of her house, a gate that is only ever closed when she goes away on holiday.

I am alone now. Some of the winter-flowering iris, of a variety whose name has long left me, which have been growing in great clumps since before I worked here, are showing a few purple flowers. These iris grow in the sun at the base of the south-facing wall where it is dry, and will tease with a few scattered blooms until they burst open en masse in March and April. The snails love the long, thin strappy leaves, and they will become brown threads as the snails, which now cluster like secretive nuts in gaps in the stone wall, cemented together by their own secretions, work their way up them from base to tip, eating along the soft green flesh between the veins and devouring the tender purple flowers. It happens every year, yet we persist with the iris, and I leave the snails where they are.

I say 'we', but in reality it is only me who does the work: all the designing and buying and planting in the garden, making all the decisions and never asking her. She never wants to know; she doesn't direct, occasionally requests. We pretend at a collaboration, as if the plants were the orchestra, I the conductor and she the audience, and that, I think, is why we use the word 'we'. We imagine that we are in control of the flowers and are going to make a show with them. A darker thought occurs; maybe we both say 'we' in deference to her. Because it is her garden and not mine. Because to say that 'I' do all the work somehow sounds rude. Very rarely, when she has seen something on television or in a magazine perhaps, she will say something like, 'Shall we plant pink tulips this year?' And of course I read it as an instruction that I should plant pink tulips. But it has been many years since that has happened.

I once knew of a gardener who talked of his employer's garden as if it were his own. His art, his craft, his knowledge, skills and vision, his labour and patience, the days and weeks and love in the seasons of the years of his life had all been used to make the planting and the various views. He carelessly used the words 'my garden', meaning 'the garden that I made', but was told with force by the new young owner that he only worked there. I wouldn't make such a mistake. This is not my garden, but it's not hers, either. Just paying for something doesn't make it yours.

Nothing is ever yours. People who work with the earth and the people who think they own bits of it see the world in totally different ways.

That gardener was put in his place by the owner, felt humiliated and walked off the job. The owner had to look for a replacement and may perhaps, over time, have learned that people other than himself had pride, too, and they were entitled to it. The gardener had a new job within days. He had to start again and take on someone else's work to make his own, as someone will take on mine, and as I took on my dead predecessor's.

Both the gardener and the owner paid for opening their mouths – silence is always good and rarely inappropriate. Words are too easy. Misunderstanding is too easy. Language is a coarse tool that fails to express delicate ideas. Words can sometimes fall from our mouths at random, without real meaning or substance, and get us into trouble. Working alone, I know more about silence than about pretty much anything else. I was taught to be silent as a child. Although there is never really silence; even the grave rustles.

It is not easy to find someone who knows about plants who will hire themselves out for the kind of money that people want to pay. Even harder to find a creative or sensitive one. Word never reached me of who – if anybody – took on that garden. Gardeners are often quiet and solitary and unused to words, but we will nod at one another as we pass, recognising another gardener in their manner and vehicles and clothes and smell. We avoid conversation, though, because gardeners are often very blunt and opinionated and easily argue and fall out with each other, refusing to speak for years because of differing opinions about how to prune a rose, or what is the best peony or some other nonsense.

Any garden belongs to everyone who sees it – it is like a book, and everybody who visits it will find different things. Not many people see this place now: the window cleaner, me, Miss Cashmere, the odd tradesman, a delivery person who can't find the front door and wanders round the back. This garden, like most others, is a trick that looks a bit like nature, but isn't really. It is written deliberately to lead the viewer into a collection of stories using colour and form, light and shade, to elicit personal emotions, to seed the imagination, to spark a journey of remembrance of forgotten things, a drift into childhood games or young loves or thoughts of people, parents, past lives, fantasies in bright open spaces or private contemplation in the shade. It is a place designed and managed to lead you on, to find and lose yourself. The way I choose to shape this or that space: wild, or tight and neat, closed or open. A scent and splash of purple as you come round the corner to a tightly hedged shady place, or flowers in blues and pinks under trees and wide-open drifts of yellow, orange and red. A bench is placed, as if by chance, where a million bees forage above in the canopy and fill that place – just that one and only place – with buzz. It is all an artifice. If it were left alone for a few months, nature's fertile beast would take over and it would become something else entirely. There are places where I let that happen, hidden from the house, where things grow wild and nature thrives. Damp spots for ferns and rotting wood, fungus and beetles, and hide-aways for hedgehogs.

Climbing Hydrangea

I'm heading back to get the big three-legged ladder from behind the sheds to start pruning the hydrangea that grows up the front of the house. I reconsider for a moment. Miss Cashmere is out, and a fall could leave me on the ground until she returned or Peggy started to worry about me as darkness came down. She would be sitting there at home, by the window writing her stories, watching neighbours walking by. Pegs would first ring my mobile perhaps and there would be no answer. She is miles away from me here in the country, there are no bus routes nearby and she does not drive. It occurs to me that I don't even know if she has the address – could she say exactly where, on the face of this planet, I've been going every day for years? Probably not. The three-legged ladder is stable on the ground and does not rock, as tripods don't. If I pay attention and step on it properly, gripping it well and not taking risks or hurrying, I can treat it as a meditation.

This is a long but simple job. On the way up, I cut off all the old flower heads, prune the hydrangea hard just above a strong pair of buds. There are dried and faded winter-bitten flowers, all the way from the ground to the bedroom windows. Decay is so often the colour of rust. Like a careless child's paintbox, all the colours mix to become the chaotic brown of the earth that gives birth to life and cosmos and colour. The shiny new buds

are rust-coloured, too. With my old red-handled secateurs I cut off the crispy flowers that fall slowly to the
ground, and earwigs and spiders scuttle away from my
hands, and snails who love their privacy simply stay
where they are, glued to the wall while they wait for the
warmth.

The weather has been getting warmer and wetter in
recent years, so the hydrangea grows faster and thinner
than it used to; tough white aerial roots of new pale-green
stems grasp firmly to the stone, tiny white hairs snaking
deep into the texture of the wall for its moisture and security. I have to pull hard to peel them away and cut them off.
I'm nervous, tugging with both cold hands, gripping the
trembling aluminium steps with my knees so hard that
my legs get bruised. Then down and back on stable land,
I rattle the ladder along a yard or so and clatter my way
back up. The pile of fluffy heads grows and, as the light is
starting to fade, I finish by raking and forking them all
into a corner, ready for barrowing down to the compost.
Tired, I haul the ladder onto my shoulder to go back to
my van and pack up. Her car crunches slowly along the
gravel drive as Miss Cashmere returns home and daylight
fades.

She is dressed all in black: a short jacket, a knee-
length skirt, tights, patent shoes with a small heel and a
square satin bow. She carries a black hat with a brim. Her
hair in her usual neat white bun. She has clearly been to a
funeral, and so I wonder about the appropriateness of
passing pleasantries with her. I ignore the outfit and carry

33

on as normal. My big smile, her small one. She is tightly drawn with a neat line, while I am fuzzy.

'Dorothy, how are you?' I ask, all upbeat. 'It's lovely to see you. Did you have a good Christmas?'

'Hello, Marc,' she answers, smiling, happy, 'it is good to see you, too – and getting on with it already. Marvellous.' She doesn't answer my question and heads for the house. 'It's nice to have you back. My daughter is coming over later. I have just been to my great-granddaughter's christening,' she says, fiddling with her door key. 'We must chat sometime.' She goes inside.

'Congratulations,' I say as she closes the door behind her. Her other, ginger cat wanders to the house, brushing itself against my legs as it goes and sits looking through the glass of the door, staring into the house as she glides away.

A Story

I'm clearing away the pile of dried flowers from yesterday. Miss Cashmere is not about, although her car is there. The bedroom curtains are drawn closed and I'm happy to have been up there yesterday pruning, because it could have been embarrassing if she were to open her curtains and see my face, or my shadow passing on them while she was in bed.

I started working for Miss Cashmere when she worked in London and she and her husband came down at weekends and Christmases, for summer holidays and birthdays and parties. Then later she stayed at home and had her three children, two boys and a girl, and I watched them grow up and leave, and she stayed. Her husband carried on going to London and coming back, and then one day in February about ten years ago he didn't return. I remember I was here, doing this same job, but not the year. He came back about a week later, for a few hours – the tide gone out of him. Cars arrived. Friends and the grown children and their new children, and his colleagues and acquaintances, some of whom I recognised, who had visited the house before, whom I had seen turning up for parties or lunches; smart people stepped down from large cars in subdued colours, and others curled themselves from smaller cars in brighter colours. The people stood around him in his box, and then they wheeled him off to the cemetery, and they made him and his box into ashes as dry and brittle as life and mixed them with the earth and sent him back to where he came from, and that was him done. He is just a story now.

I had asked Miss Cashmere if she would rather that I didn't come to work that day, but she said, 'Carry on as normal, I'll be pleased to see you.' Maybe she wanted the normality of me pottering around. So for me and the plants and the insects, life went on, as the people grouped in black on the other side of the steamed-up conservatory windows stood around in their glass box among the imported hothouse lilies, drinking briefly from small-stemmed, dark-filled glasses – sherry perhaps. Outside, I kept my distance and worked near the stables and the compost heaps for the warmth. I have spent more time in his garden than he did. He was a nice man; we spoke occasionally, and he was friendly. We were different. We enjoyed the same whisky. He was clean and neat and polished; I am none of those things. We came from different worlds. We believed different things.

Now he has been dispersed into nature. He believed that flesh was sinful and, because of this, we struggled to communicate about purpose and meaning, but simple subjects like love is good and beauty is good, and work is tiring and whisky is good, were easy. There were a million rules and beliefs and systems and rituals built between him and me and the truth of mud. He lived for a higher ideal, a great idea that, for him, gave life a meaning. He was raised to be special. I was raised to be nothing, but I've tried very hard to make being nothing into a good thing.

She had loved him and was sad for a long time and, as far as I knew, never took to another man. Miss Cashmere is alone. Sometimes the children come, mostly they don't. I am here every day. I let myself in, do my work, then drive home to Peggy, whom I love. Love is simple: just pay attention, put the effort in, kill your ego. Peggy does the same, so it works.

Cyclops

There are primroses flowering, pink and yellow. Some of the hellebores – Christmas roses – are showing their white flowers. In Latin, the Christmas rose is *Helleborus niger* (the black hellebore), although the flower is white – the root is black. Hanging face-down in a cool shady place, in the drying leaves under the shrubs, where they like to be. When they fade, their seed pods will swell and turn brown, and their relatives the Lenten roses (*Helleborus orientalis*) will open their red-wine flowers. Hellebores are common garden plants that are deeply poisonous, like many other plants in the garden: the foxgloves and aconites and rhododendrons, and more. An unrelated plant, known as the 'false hellebore', contains a poison called cyclopamine. The child of the pregnant woman, the cat, the goat, the chicken that consumes the plant is born with one eye in the middle of its forehead and a primitive smooth brain, like a snake or a mole. The creature dies soon afterwards. There is no recorded instance of a Cyclops living into adulthood.

Helleborus leaves tend to turn black when the flower blooms; I kneel and take the secateurs from the cracking leather holster that hangs on my belt and cut them off, so they can display their flowers the better. It is a traditional thing to do; the plant doesn't need them any more, it is about to go to sleep. The sap, too, is poisonous, but it has never had an effect on me, though who can tell which is cause and which is effect, of all the daily thoughts and feelings, aches and pains, tinnitus and odd heart rhythms? Miss Cashmere probably won't even notice the blooms, and I complete these small tasks for my own pleasure now, as much as for hers.

In 'Warming Her Pearls' the poet Carol Ann Duffy writes about the traditional practice of a servant warming her mistress's pearls before she goes out for the evening – a sensual poem about a relationship that has personal value to the servant, but only a functional one to her mistress. I wonder if I'm warming Miss Cashmere's pearls. I love her as I love the poisonous hellebores and the faded wintersweet. She is a flower in my garden, and I have wondered what she felt about me. We have been together for so long, perhaps it gives her pleasure to have me around; or, more likely, my work here only gives her one less thing to worry about. I think that in years gone by I was an entertainment, seeing her and her female friends drinking and laughing with each other in the conservatory or on the patio in the sun, as I worked in shorts, mowing the grass or pruning the roses. I was young and fantasised about being invited inside, and what might happen. But of course I never was.

Eventually it occurred to me that she was just watching me working; she had no intention towards me at all. I imagined that perhaps she was trying to understand what I was doing, what my reasoning was when I pruned some branches of the apple trees and not others, or why I was digging in a particular spot, what I was planting for her surprise and delight. But then after a while, as there were no questions, I realised that the way she looked and smiled at me was exactly the same as the way she looked at the flowers, or her lovely house or new Jaguar sports car, or an expensive vase on her table. It was pride of ownership. Her thoughts and feelings about me were the same feelings she might have for anything else in the garden – the flowers, the birds and insects, I belonged with them. I was a hired character on a stage.

Any garden needs a gardener to look after it, preferably one who looks like a gardener and who can be a decorative part of the picture. In the eighteenth century there was a fashion for ornamental garden hermits, who would live permanently in a purpose-built hermitage on the landowner's estate for the entertainment of the British aristocracy and their guests, who would seek out the hermits' advice or watch them, for their own enjoyment. She watches, yet she is entertainment for me also; each of us completes the garden for the other.

Code-breaker

She doesn't come out much when it's cold. She used to be very beautiful; her ankles have become skinny, making her feet look long, her thick brown tights the colour of a latte, she is bent over and doesn't say a lot; her eyesight is fading, she seems smaller.

When I rarely speak of Miss Cashmere to others, that is what I call her, that is how she is known, but her name is Dorothy, and that is what I call her when I speak with her. When I speak to Peggy about her, I call her 'Dotty', although she is far from dotty. Old people have a strength and a frailty that the young do not have – well, a different kind of strength; a frailty that is different from that of the young. She is weak physically, but her mind, and her spirit, her self-knowledge, are strong. The young are the opposite: their bodies are strong, but their self-knowledge, their calm, is weak, and this is the natural way of things. My calm is strong; it is the strongest thing I have. I've nurtured it as a precious thing for many years.

She smokes about forty cigarettes a day. I asked her once how she was so fit and healthy, even though she smoked so much, and she said, 'Do you know, Marc, I have never done a day's work in my life. I am useless – useful things get used up and worn out.' She said, 'The useless ones get left completely alone on a shelf. Like the royal family or a cracked teapot, I am totally useless and so, like them, I will go on for ever. I am purely decorative.'

She smiled that lovely old-lady smile, which made me smile too, and then walked on, bent over, still smiling, cigarette in hand. I knew, however, that this was not true. I knew her when she worked. I have been told that she worked for the government, doing something that she would not talk about. Everybody knew. Nobody knew what it was that she did, but those who speculated on such things had ideas they were only too willing to share: 'She was a code-breaker' . . . 'She was a spy in Russia or Norway in the Cold War' . . . 'She was a spy for the government in the BBC during the Sixties, when television was full of working-class people, playwrights and left-wing film-makers who wanted to upset the apple cart and needed an eye kept on them' . . . 'She worked at a listening station'. She was always some kind of spy, in most people's imagination, but nobody knows. People like stories because they are easier to deal with than the unknown.

She never told me what she did; her husband had been a barrister. He told me, because I asked him; I don't think I ever asked her what she did. Very early in our relationship it became clear that a straight answer would not be headed my way. She was a master of misdirection, and that made me think the gossips were probably correct.

Wood Pigeon

I'm making porridge for breakfast. The wood pigeon makes his soft five-syllable call over and over at the end of the darkness. A call that all country people know, and some describe as 'My-toe-hurts-Betty . . . my-toe-hurts-Betty . . . my-toe-hurts-Betty.' The first line of a haiku, over and over again. I want to rewrite the second and third line and complete the verse, but lose hold of the thought and go to make tea. From the kitchen I can see him or her, solitary on the television aerial of the house behind. He has been silent all winter, but spring is coming and he has begun to call. Last spring there were two of them. The haiku comes to me on its own, as they often do when I have planted the idea and then turned away:

My toe hurts, Betty
and where have you gone, my love
I'm waiting for you.

I tell Peggy my wood-pigeon haiku and she says that the first line should be 'My-soul-hurts-Betty' or 'For-God's-sake-Peggy'. I say, 'Oh, for God's sake, Peggy' and chase her to tickle her under the arms, and she runs away screaming and laughing and then, with a fake-serious expression, says, 'Don't you dare!' and picks up a spatula from the counter top to threaten me with. Peggy is usually happy, but she can be dark sometimes. I grab her and cuddle her hard and she wriggles. The cat watches us from her curled-up place on the chair; she looks jealous sometimes, so I pick her up and cuddle her too, and Peggy laughs at the way the cat just abandons herself, lies back in my arms with all four legs in the air and looks into my eyes.

'She adores you, that cat,' says Peggy.

'And so she should,' I say.

It is cold and frosty, with a big three-quarter moon low in the clear sky. From the kitchen window there is a pine tree silhouetted against it. Closer, a male blackbird is quiet at the topmost point of the bare lilac tree. I am watching to see his beak open and hear his call arrive across the gap between us. Around him, new leaf buds have appeared on the lilac. The other birds are silent, waiting for the day to warm their prehistorically cold blood. I eat breakfast, and slowly, but just fast enough to see, the moon slips behind the house and the full daylight is here as I drive to work.

The buds on the big-flowered magnolia (*Magnolia grandiflora*) are tight; a million candles in khaki velvet-green are ready to burst open, turn pink and white and fill the trees with light. Little clumps of snowdrops (*Galanthus nivalis*) sprout through the grass. Dangling chains of pale-yellow flowers hang from the neat, arching bronze branches of the stachyurus (*Stachyurus praecox*). There are winter-flowering cherry blossom and pink viburnum flowers on bare branches. Camellias' waxy flowers are out in red and white against their hard, shiny leaves.

There is always something to do. This place needs another gardener really, but I guess there isn't enough money. Most of my jobs are routine and, through such routines, life becomes calm. In February it is time to start tidying up the plants ready for the spring. Today the ground is frozen – the puddles of water here and there are thick ice. I'm working on the climbing and rambling roses that grow on the wall at the front of the house. All the pruning needs to be done before the soil warms and buds start to break, and the sparrows collect twigs and the blue tits and other small flocking birds collect moss and build their nests in the unpruned rose thicket and ivy. Some of the roses grow over the wall and into the road and need cutting back a little; some have been brought down by winter winds. There is a flower bed in front of these old plants, and I have to walk on it to get to them. I usually avoid standing on and compacting the bare soil that I spend so much effort on improving – there are perennials waiting under the earth here in front of the roses: lupins and gladioli and tulip bulbs – but the soil is hard and frozen, so it is okay to step on. There are small green buds on the roses already, as the winter has been mild again.

My hands are cold, and the warm scent of the string I'm using to tie the branches in feels comforting, like the unravelling of a scratchy wool blanket. Climbing roses do not need too much pruning, just tidying up. A thorn digs deep and I bleed. I've left my blood in many places. On tools and bed sheets and people and flowers. My DNA is now part of the air we breathe. Violence is never far away for a working man like me; our blood runs close to the surface and comes out easily. It is to be ignored. Sometimes, by some, its release from the body celebrated as a sign of masculinity.

The Old North

Men of my age in the Old North were all brought up in much the same way. Educated to be at war, ready to fight with a foreign power when needed and, when not needed for that, we had football or rugby and hard physical labour to keep us fit and ready, until we were worn out and replaced by younger versions of ourselves. We were useful, so we didn't last. Our value was as a mass, not as individuals. Nature doesn't require individuals; they are disposable as long as there are enough of them. School taught us to withstand abuse, boredom and pain, and we learned that life was all about hierarchies and power.

In those days everybody had a job up there. People worked in the same kind of jobs as everybody else in their community. I could work down the pit, like the men who used my father's pub. I could join the army or work with steel. The girls could work in the bleach works, or become hairdressers or cooks, or work in the cotton mill. My mother was a cook. Her mother had worked in the cotton mills. The mill girls were raucous and what my mother called 'common', and I would have to run away from them to hide when I was an apprentice, because they would strip a boy's clothes off and make him run around the mill naked while they chased him and laughed. They would confuse a young man with their lascivious behaviour, and would slap and pinch him to make him run. They were famous for it.

When I first arrived in that cold northern coal town I fell in love with the women who worked in the bleach works. They walked chatting with their friends, about twenty of them, clomping along the flagstone streets on wooden-soled shoes from the nearby cottages where they lived, and in through the big green factory doors that rolled along a track on squeaking metal wheels. They entered the darkness of the red-brick shed the size of a rugby field. The bleach works sat in a valley over the canal, which contained the smell that could still burn the nose half a mile away downstream. They tried to keep the smell behind the doors, but it leaked out. As the women walked they seemed close, serious and yet comfortable in each other's company, talking quietly; sometimes a laugh, but mostly they were quiet, unlike the mill girls. Their eyes shone out from scrubbed, almost pure-white skin, pure-white hair, white eyebrows and lashes, white coats, white dresses, trousers, tights, headscarves and shoes – all white from the fumes that rose from the vast baths of hot bleach that I imagined they worked with. Even the irises of their eyes seemed washed and pale, like old denim jeans, and I fell in love each time I saw them walking proudly, elegantly, into the works on the edge of town by the canal.

They looked like they were all from the same family: sisters, aunties, mothers, daughters, mothers-in-law. Probably many of them were. I thought they were a large family of albinos. They were pink fairies that I used to seek out, just to look at them. I wanted to see them naked. I was beginning to get sexual urges that I didn't yet understand. What were they like, under their work clothes? Were there colours under there, or was all of their clothing bleached white, like the skin? What colour was their underwear? At first these were not sexual thoughts, but curiosity about their ghostly, seductive whiteness. But as I thought more about it, I imagined their bodies and the images turned sexual. Later I learned about their monthly blood, and I imagined that to be the only colourful thing about them. Red streaks and clots coming from the middle of empty, pale-white ghosts. Ectoplasm. Thick and red and tying them to the earth, so they wouldn't float away.

I stared at them and fell in love with them over and over again – sometimes with this younger one or that older one, because of a pigeon-toed walk perhaps or an elegant neck, or frizzy hair, or occasionally a sniffy red nose from a cold or the fumes. I would look for the one that I saw the last time, but I could never identify her; they were too alike. They ignored me, a child. Some of the younger ones, not much older than me (fifteen perhaps) had just left school; the bleach worked fast, and even they were white from head to toe. The older ones who didn't work any more – the retired grandmothers who came with prams, so mums could see their babies in the lunch breaks – looked fresh-faced, smooth-skinned and pinker than their daughters, appearing decades younger than their years, often with tinted hair, blue or purple rinses. You never saw really old women then, not round there, and no really old men, either. They died young. Like the miners, coughing their lungs out, bubbling pink blood onto white handkerchiefs. And those babies came in prams to see where they would work when they grew up.

The miners in the pub, drinking beer after work, playing darts and dominoes, all seemed to be wearing black eyeliner in the creases of their eyes, made-up like my mother. I couldn't help but look at them. I stared. They became offended and asked my father what was wrong with me. Already an outsider, not a team player, this confirmed their thoughts of me as an oddity. I liked to draw, and drew some of them and the pictures were hung up on the pub wall. They asked my father if I was queer. I was a writer and a reader and spent my time alone, and looked into their eyes instead of playing football and screaming with the other boys. I thought they were wonderful; they excited me and I didn't know why. I think now that what excited me was transgression. Later I was told, pushed up against a wall by one of them, that if a man looks into your eyes, it means that he either wants to fuck you or kill you: 'Which is it, Sonny?'

His world of black and white appalled me: the win and lose, us and them. Devoid of shade or colour. It was not safe to be anywhere outside that simple duality. I grew to detest duality. I always want every game to end in a tie. I fear and mistrust anybody who expresses any kind of certainty about anything at all. Much later I learned that it was okay to enjoy pretty things, but as a male in that particular world the list of things I was allowed to enjoy was very restricted – either sentimental or concerned with power and competition – and I was not to use the word 'pretty' to describe them. I could enjoy flowers, for instance, dahlias or roses, but only if I grew them competitively, aiming for a standard that had been devised by a committee of some sort, and entered them in shows to win medals. Conformity was key. Certainty written down in books that clearly defined what was a win and what was not. Prettiness or a vague liking 'just because' was not an allowed part of the process. Clarity was everything – everything I grew to despise. Win or lose was all that mattered.

My father told me that when the men came out of the mine at the end of the day, they would be covered from head to toe in coal dust and would have a shower in the pit-head baths before they left work. When they show-ered, the coal dust would remain lodged in the creases of their eyes. The black coal eyeliner framed the pale blue of their eyes. They were all pale blue round there. My eyes are pale blue.

Before I went to that pit town I lived at the seaside town of Blackpool: we moved house a lot. My people were always travelling. There was no home, no lasting friend-ships. We just lived wherever we were, and where we were was where work was. When I was around eight there was a boy about my age who went to a different school from me; his back yard faced my back yard. On a sunny day he would come home, go inside and come out to play with us, wearing a dress. Borrowed, I guessed, from his sister. I used to wait for him. Leaning on the wall, like a Blackpool gangster waiting for his girl. I didn't get changed. I only had my school uniform. The boy would ride his bike up the back alley towards the rock factory at the end, which filled the street with the smell of cooking sugar and peppermint. I would run alongside, racing with him. He would pedal like mad, standing up on the pedals, his skirt flying and his bare legs flashing. He wore white sandals. I was a good runner. He had the only bike in the street and I wasn't allowed to ride it, his mum said; nobody was but him. I envied him his bike and his pretty dress. If I had dressed like that, my father would have battered me. He told me again and again not to play with 'that queer kid'. 'You know what I'm talking about,' he said, when I asked him why. I didn't know, but in a way I did. It didn't stop me playing with him. I loved him and was excited by him, thought of him as my girlfriend. Transgression. We were eight. I never saw anybody like that when I moved further north later. He was somebody's flower. I wanted to be somebody's flower, too.

I left school at fifteen in that northern town and worked as an apprentice, making immense boilers for factories where coal or oil would be burned to make high-pressure steam to power heating systems or large engines. I learned to drive a crane, moved massive steel plates around with a hoist, drilling, grinding and welding them together. We drank competitively and played the bucket game. The game was to see how long you could

52

bear to stand, holding a steel bucket on your head, while your colleagues beat it to a dented mess with whatever they could find – iron bars, hammers. We set fire to passing workmates with welding torches; welded people's tools to their benches; waited for a worker to go to the toilet, then set a trap with an arc welding set, putting one end of the circuit in a puddle and attaching the other end to a bench where a spanner or other metal tool was waiting for them to pick up. We threw welding rods at rats coming in from the canal just outside. We prepared for war and ended up with tinnitus or post-traumatic stress disorder.

At the tea and lunch breaks some of the men played darts or read *The Sun* newspaper or looked at pornographic magazines. I broke ranks and at weekends wandered into a bookshop in town and, without guidance, bought books that I thought might be important. I had never been in a bookshop before I earned my own money. I didn't know what I should be reading, so I read serious-looking books and whatever poetry I could get my hands on – usually American, as the old English poets I detested in school wrote as if nobody other than the classically educated were worthy of poetry. The first poet I ever read from choice, Rod McKuen, talked to me in 'Stanyan Street'; he introduced me to Jacques Brel and his song 'Amsterdam', and I knew from the first time I heard this that I was European.

I knew nothing, so I read everything that looked like it might be important – books with covers that didn't feature cowboys, gangsters, spaceships or semi-naked girls – and while stabbing in the dark, the world stopped being black and white, I lifted up my nose and scented a sweeter life. The men around me thought I was looking down it, and pushed me further out. I had no interest in the things that I was expected and trained to think about: pop music, celebrities, football teams, politicians, the Catholics or Protestants, foreigners, the Royals. I was fascinated by girls, but they were not interested in me. I was a weirdo who read strange foreign books and didn't play with the boys. As far as I was concerned, the boys didn't play very nicely, talked rubbish and hate, and made me feel like killing somebody. I believe that was what I was supposed to feel.

In 1974, when I was sixteen, my mother choked to death in the night after a long illness, her lungs ruined. Not long after the funeral my father told me to leave, I was too odd for him, and so in the course of a few weeks I lost my mother, my family, my home. Unable to sustain a job while living rough, I lost my income, too, and became a tramp, a gentleman of the road. I was reborn as 'The Fool' in the Tarot who, in flamboyant clothes with a dog snapping at his heels and sniffing the scent of a flower in his hand, looks only at the sky and sets out without a plan or a care, or any awareness of the precipice in front of him. I still like flamboyant clothes: I'll put on a bow tie at the drop of a fedora. I walked for hundreds of miles, day in, day out, miserable and happy, hot and cold, wet and dry, and in the journey the last threads of any sense of self I ever had blew away in the breeze. Every minute was a miracle and, like Camus, I became happy just to be conscious.

'I'm here, are you there?'

Miss Cashmere locks her kitchen door and totters over to her car. The sparrows are singing, calling out to each other in the roses: 'I'm here, are you there? Yes, I'm here, are you here, too?' It has rained a little and they sound joyful. When she goes out she is impeccable, precise, and I stop working to watch her. I like to see people dress as if they had thought about it. Pale and wispy hair still with a hint of ginger in it, tied up where it used to be long and red, and often in a plait. She is in a green skirt and jacket, smart dark-green coat, startling burnt-orange tights, flat brown shoes and a flowery headscarf printed with roses perhaps, on an orange ground that chimes with her tights – I can't tell from this distance, maybe they are peonies. I have seen her many different styles over the years and think of her as generous for adding brightness to a sometimes drab world. She turns away, folds herself into her car.

When I was a young man I admired those older men who always wore a tie, who were gentle and polite, who said please and thank you, who opened doors for others, who smiled at shopkeepers and passers-by. So as time passed, I tried to become one of them, because they are thoughtful and add colour and loveliness and I, too, could become a flower. I started saving up for good shoes, and wearing a knitted tie under my tweed work jacket. I still do, if it is not too hot, but I am used to the cold and get hot easily.

A pheasant bursts out from its scrape, crarking and flapping like a plastic bag in a gale; it flies hard and fast towards the bottom of the garden, over the low beech, hazel and alder into the field, and is gone, down to shelter. They always startle me, camouflaged, invisible against the scrub in the shadows and waiting until I almost stand on them before they take off. On the other hand, the foxes in the evening do not hide; they watch me and I watch them. They are predators and are not afraid, but slink off low to the ground if we are too close to each other.

She drives away in her ancient oil-green Jaguar, slinking off low to the ground, clouds of exhaust in the freezing cold air. She didn't notice me in my brown corduroy trousers and a green moleskin shirt and brown tie under my old tweed jacket. I, too, am probably invisible, camouflaged like prey or predator. I have been here so long I am part of the background. A clear blue sky now, super-blue, ultra-marine; there is some kind of hawk crossing, maybe he saw the pheasant and is on his tail. I do not know the names of hawks or their shapes. I never learned them and don't need to know what they are called. They are too far away, and my eyesight is not good.

In my sixty-third year on this planet I go to work with a pruning saw and lopping shears by the wild hedgerows at the bottom of the garden, beyond the trees and in the wilderness. Cutting back the brambles and tidying the branches of the shrubs and blackthorns and hawthorns and the stunted battered ash that form the boundary. Even after all these years I spot a likely place in the shade where the frosted leaves are dry and deep – they would generously give shelter and be a good place for a vagrant like me to sleep for a night. Cutting into a hawthorn that has a long, cracked branch coming over the broken wooden fence into the garden releases a scent: powerful, like roasting chestnuts, then fading to a weak varnish smell, a perfume that's an invisible cloud fading downwind. A memory of walking to the shops behind my mother in her tweed coat, her hair in a headscarf, gloves on her hands, high heels. Her favourite perfume, also called 'Tweed', which left a faint trail that followed her, as I did. She carried her wicker shopping basket and bought vegetables that smelled of mud from the greengrocer, who smelled of carrots and cauliflower. I was in blue short trousers and sandals. The memory is all the more delightful because it is incomplete, a fragment that begins with a scent and ends with the sandals. There is no more to it.

The magnolia buds are starting to open, a million, a billion of them, and the roses are sending new stems out. The mahonia (*Mahonia japonica*) by the wall is flowering, little spikes clustered with tiny yellow flowers that have a sweet milky-balsam smell. It has been in flower since autumn and will continue for a little while yet. On the sunny bank beyond the trees there is a massive cluster of daffodils, all perfectly open and blooming early on this cold winter day; and the cotoneaster tree is dropping its berries, which have lasted since November, on the path and staining it red. Two military planes fly high, leaving perfectly parallel lines from one horizon halfway across the vast sky as they head for the other; the pilots who sit inside, highly trained and helmeted, skilled with joysticks, play at slaughter. They hardly seem to be moving, from this distance. They are rapid, but the world is vast. At my feet, by my right boot with its frayed lace, a perfect winter aconite (*Eranthis hyemalis*) flowers; a tiny winged creature, a fly of some sort, crawls across its simple waxy yellow petal.

The winter aconite is related to the buttercup and looks a little like one. Also related to the buttercup, and also known as aconite, is wolfsbane (*Aconitum napellus*); this is a deadly poison – even contact with the skin can cause numbness and tingles, and heart problems for some. Wolfsbane has a lovely purple flower and looks nothing like a buttercup, more like a delphinium or larkspur, and it grows in the wild garden. In fairy stories, wolfsbane (also known as monkshood, because of the shape of the flowers) is used to destroy werewolves: a touch from the flower causes them to fall dead at your feet and turn back into the human form they once possessed. It's said that wolfsbane was used to kill troublesome wolves – raw meat was spiked and left out for them. Apparently government spies in Nazi Germany also used it to poison their bullets.

The buttercup is a poisonous family of plants (*Ranunculaceae*), as is the nightshade family (*Solanaceae*), which includes the potato, tomato and aubergine as well as black nightshade (*Solanum nigrum*) and the decorative but poisonous Chilean potato vine (*Solanum crispum*), which grows over the top of a wooden fence further down the garden. I am surrounded in this place by poison and myth, and imagine firing an arrow tipped with wolfsbane into the aircraft, destroying the helmeted wolves and turning them back into human beings as they fall to Earth.

As I leave in the late afternoon there is a frosting sky, clear as ice, rusting straw-coloured. I can feel tiny, sharp ice crystals falling on my face as the cold sun sets, glittering in the rough weave of my wool jacket. Daphne blossom on bare stems is still pink in the last of the day's light. Lots of late-winter flowers: their names are poetry, forsythia, *Skimmia japonica, Hamamelis, Cornus mas.*

She Needs a Stick

Every year somebody gives Miss Cashmere a skimmia in a pot. *Skimmia japonica* has become a traditional Christmas plant. Its deep-red buds and shiny green leaves give it the proper seasonal colours. After the holiday is finished, she brings it outside to where it really belongs and, over the years, I have planted them in various places. They are all looking good; they rarely grow much over three feet tall and wide and do not spread too much. They are no trouble and grow well in the shade, where few other flowering things will grow, and the tight crimson buds will soon open into massed white flowers. There are so many skimmias here now, yet here she comes with it in her hand, smiling and asking, 'Where can it go?'

Her skin glows pale, her lips pale, her eyes pale; she wears her rose-print scarf and green coat, and she is all colour and brightness and pale. My first thought is to answer, 'On the compost', but of course I don't. I tell her, 'I'll find a place for it behind the summerhouse, where you'll be able to see it from the window. I'll leave it there in its pot.' I say, 'To plant when the soil warms up.'

She smiles and says, 'Perfect', and just as I am about to ask her if she is doing anything nice today, she turns and hobbles slowly away, leaning forward. I can see that she needs a stick.

March

Grass Sprouts, Trees Bud

Wind and rain carve spirals into my skin, and birds blow by with beaks full of twigs and it feels like spring. Everything is turning. Slowly the warming sun is bringing the acres back to life, and my arms and back are growing strong through the aching toil of digging and moving compost. Work makes muscles strong, and my focus on this simple task makes my mind accept what my body has to do.

I'm moving compost from the heaps by the green-house, digging into the fresh-smelling mountains of sweet black earth that I built and turned last year. I throw nine heavy forkfuls into the barrow until it crests the top, then push the load across an acre or more of lawn, tip it out and spread it around the roses, and go back for more loads to spread where the lupins and gladioli crowns are showing signs of green; then six more barrows to where the bank of red tulip bulbs are buried; then another five on the bed where I'll plant the dahlias and, as I count, I wonder why I'm counting. There could be only two reasons: either it is so that I can complain about my life or so that I can brag about how hard I've worked – and there isn't anybody to brag or complain to. I stop counting. There is no earthly use in knowing.

There is a pain in my left knee, deep, sharp and sweet, the result of kneeling, lifting, twisting, cutting or tearing something. Hard labour hurts, but pain is just another sensation, one of many; like hunger, love or sex. I decide to focus on breathing in and out into the peace, into the silence, into the rhythm of breath and tread and flow, until there is birdsong, the squeaky wheel of the heavy load, the clear sky, cool air, the million scents and textures under-foot, and nothing but awareness of each passing step. Sisyphus rolling his stone.

Last year this job took a whole day and at the end I was exhausted, but over to the east there is cloud coming that will bring rain later, so I'll probably not be able to finish this today and could be taking it easy by lunchtime. Had I continued counting and rehearsing my complaints, my head would have been full of waste and I would not have noticed the coming rain, and so the day would have seemed much harder, darker and longer than it really is.

As I approach the house with yet another squeaky load, Miss Cashmere hangs out of an upstairs window and asks, 'Is it cold? Will I need a coat?' My body is hot, my shirt wet with sweat. I live and work outside in frost and rain, wind and heat; the way I feel about the sensations on my skin and in my body is vastly different from hers. I have been homeless and lived outdoors, day and night, summer and winter. Awareness of the vast chasm between our experiences of the world floods in. I know a bitter depth of cold and chilblains and hunger that perhaps she doesn't. I grew up with a small coal fire – when there was coal available – and just one set of clothing. What she feels as cold I might not even notice. I am unsure what to tell her. Is it cold? Will she need a coat? I have no idea. It's too much responsibility. She is as deaf to my predicament as a baby would be and only wants to know the answer to her question. I tell her that she will need an umbrella. That appears to satisfy. We live in the west; an umbrella is always advisable.

The rain comes shortly afterwards, and I push the empty barrow deliberately slowly through the shower to cool down. I take shelter in the warm, dry greenhouse. A fearless pheasant follows me to the door, looks at my lunch, hangs around my old canvas bag. Somebody must have been feeding him, because he is almost tame and so he will not survive the shooters when they come in October. The tortoiseshell cat sits and watches with calm intent and a focus that is deeply admirable. She flows along, completely in the moment, with full awareness of her complicated body. Nevertheless, the pheasant becomes aware of her moving and blasts off, leaving the cat looking wide-eyed at the bird's disappearing backside.

A big black umbrella approaches slowly, with Miss Cashmere hanging almost weightless underneath. A dark dandelion seed drifting over the wet grass – an arm, a thin body, two skinny legs all dangling – dragged along by the slight breeze, slower than the slow-moving air, hovering just above the ground, where she might land and take root at any moment and flower. Heading to the summerhouse, she floats by, smoking her cigarette. As she gets closer she

65

becomes heavier and more solid. Sheltering under her umbrella, struggling with her newspaper, her cigarette packet and lighter and lit cigarette, each of them as important to her as the others. Hanging on to them all.

She doesn't come into the greenhouse. Hair still tied up, but no flower today. Green speckled jumper that looks too big (has it stretched or has she shrunk?), red flowered scarf, long green woven skirt that matches her jumper, green wellies, brown woolly socks peeping over the top. I take her in. I watch her go by, trying not to be seen to be watching. She dresses from the earth, a bit old-fashioned, in clothes made of wool and silk and cotton that were expensive thirty years ago and have far outlasted their initial cost. She appears weighted by the things she carries, old dark things. As if they pull her down towards the earth and, without them, she would float up and away for ever. As she reaches me, she has landed and plods doggedly to the summerhouse.

Cosmos

In the greenhouse I am planting cosmos seeds (*Cosmos bipinnatus*). I sow great banks of them every year. I put a handful of a rough mixture of compost and sand into two-inch pots and then in each pot, using my finger as a dibber, I make a little hollow in the soil and put just one tiny black crescent-shaped seed, looking like a sliver of a toddler's dirty fingernail, into the depression and crumble a thin layer of compost over it from between my fingers. I sow three packets of seeds, and arrange the 450 square black plastic pots, which will be plastic for ever, on the old cracked wooden benches that are silvery-grey and brown and jewelled with emerald lichens in their cracks, and smell of damp and soil and fertiliser and years of gardening. I water them from a galvanised can that lives only in the corner of the greenhouse, because it is too old and rusty to be carried very far and in a few years will be scrap. Shiny clear water dribbles from the rusty spout into the pots, and old and new blend in a timeless ritual that keeps me, the intermediary, in employment.

My hands are dirty; there is blackness under my nails, grime in the threads and coils of my cracked skin, around the shiny calluses on the hard pads of my fingers. My fingers are turning slowly into paws. I have learned how to be dirty. There is clean dirt and dirty dirt; the dirty dirt stinks of shit and disease and decay and gets washed off, no matter how cold the water in the outside tap is. Clean dirt has finished its rotting process and has become wonderful, wild living earth; it gets left on my hands. Nevertheless, it is alive with fungus and bacteria, teeming. It smells of life. It is my biome.

Over the weeks nearly all these little crescent seeds will send out two tiny feathery leaves, and from between them a stem will grow straight up and send out more ferny leaves. Like every living thing, they will flow from nothing – to being and back to nothing again. They happen of their own accord in their own way, and I cannot mould them or force them to be what they are not; that would break them, or I would break trying. I let them flow and become their strongest thing, like raising children. They do not have long, so all I can do is give them a happy start. These simple plants taught me these basic truths of how to be, with myself and others.

When the soil outside has warmed enough to dig into, and the last frost has passed and the plants are nine or ten inches tall, the pots will be full of coiling white roots. I'll kneel and plant them out. I could instead wait a few weeks and scatter the seed straight onto the icy ground; they would germinate as the soil eventually warmed and would be fine for a while, but slugs and snails would eat the first tender little leaves and not much would be left. They are my children, so I want them to be strong enough to thrive on their own. I don't mind the slugs and snails, I do not poison them. I avoid chemistry in the garden, and the invertebrates are part of the way that soil is made and waste is recycled, and the earth needs them doing their work, but I want the plants. What the earth needs is far more important than what I want, and I refuse to attempt to fight nature. I sow the cosmos seeds in pots and plant them out when they are strong; the slugs will get a few of the lower leaves, but the plants will be full of energy and the flowers will grow – wants and needs will be in harmony.

In ancient Greece the word *kosmos* was used to describe the orderly and harmonious universe, the opposite of *khaos*, the primeval void that came before. *Khaos* was made from the four elements – earth, air, fire and water – a vital energetic place where everything fell endlessly in all directions, as there was no solidity, no up or down, no earth or sky, a cauldron of creativity that gave birth to gods. Those same four elements will work together to produce the cosmos flowers that I sow. I'm playing with chaos: we dance.

It will take the rest of the day to sow these seeds. I momentarily wish I had a chair in the greenhouse and could sit while I do this. I remember having the same thought last year. I stand at the bench and, when I am hungry, go and sit in my old van to drink a flask of tea and eat the last piece of the Christmas cake that Peggy baked for me, while I shelter from the light but persistent rain. Peggy's Christmas cake is full of brandy and would last for years, if I wasn't around.

I'll plant a multitude of cosmos and make a gently swaying thing that will give me pleasure and perhaps, I hope, her too. An individual flower, and the colour and the pattern of a growing moving plant, can absorb the attention – well, my attention – for hours anyway (I am easily distracted by something pretty). But cosmos should never be alone; they are orderly and peaceful and deserve to be in company. Five hundred plants could make up to 2,000 flowers and become a tribe, a flock, a shoal that switches this way and that in harmony. The massed daisy-like heads of pink and purple floating three feet in the air, swaying and mapping the invisible breeze, as varying air pressures try to stabilise.

Cosmos thrive in poor soil, and as long as it rains from time to time and they get sun, they can be neglected; perhaps a little deadheading as the flowers fade, just to keep them pretty, make more blooms. They are simple annuals and will die at the end of summer. I'll cut down the lifeless stalks in the autumn or the spring.

If the cottage garden at the front of the house is a Monet, with its lupins and larkspurs and roses and apple trees, foxgloves and bees, then the immense back garden is a Rothko, or a Miró, with, in midsummer, great patches of colour contained by line. A curved swathe of pinks and violets here, then a round bed of bright-red roses, each patch enclosed in a field of brightest green. I am a painter with flowers. Of course it changes all the time, but this part of the garden was designed to peak in midsummer, when Miss Cashmere was young and there were parties. It still peaks in midsummer, but now because it is old and fixed in its ways.

Perhaps I should dig a new island bed in the lawn and make a splurge of colour – white possibly. Destroy some sadness. I do not love the lawn. I pity the controlled and stunted grasses that are cut down again and again and never allowed to seed themselves; there is no life, no growth there, no renewal or chaos. A sterile green field designed only to be used as empty space between the changing colours. The lawn is relatively unchanging while the seasons come and go: its height alters a little, slowly as it grows, quickly when I cut it; sometimes it goes yellow for a short while when we have a drought. If I go away for a day or two, or can't mow it because of rain, the irrepressible dandelions (my favourite plant) appear above the grass – they only take a day to flower. Life is change. The lawn's job is never to change, to remain immortally green and flat.

Other species are creeping into the lawn – different kinds of grass, vetches and daisies and dandelions, nettles and thistles and clovers – and I imagine it left for a month or a year. A dream of locking this garden up, reclining in a big chair, like a Canute washed over by green waves, drinking a perpetual mug of tea. I would end at its edge as it flowered and meadowed; I'd be wrapped in spider silk and stuck all over with fallen leaves and grass, then brambles would come and wind around me, and blackbirds and thrushes nest in them, and hedgehogs in a ball of leaves under my chair, and mice in my boots around the tiny bones of my feet. A robin nesting in my hanging jaw.

The early nightfall comes and I am tired. A good tired, the tired of a day's hard work that will give me deep and peaceful sleep and pleasant dreams and let me arrive at the new day satisfied.

March Frost

The blackbird, always the first to sing each day, from his usual place at the top of the pine tree, goes silent when I come outside. There has been a frost in the night and I start the van and scrape the windscreen. When I get to the garden there is more birdsong: blackbirds, robins.

The shadow of the house is thrown on the ground by the low sun and, instead of being dark, the shape is white with frost, a negative image. As the sun moves through the day, the shadow moves around, hiding behind the building. A monochrome image of a shrinking and ever more distorted house is cast on the earth, and as more frost is exposed to the sunlight, it melts until the grass is glittered with sparkling drops of water, sequins that burn off in the warmth to the green of new grass growing. A slowly developing photograph in a dish. Her birdcage, her cosy nest, her prison. Where there is no direct sun, at the base of shaded walls and hedges, the white ice crystals remain all day. My dark footprints, the hooves of a beast, go across it. My bare worker's arms, tattooed with Celtic swirls and a stag's head – blue-inked songs that are carved, unchanging, on my living skin and remind me of my journey – are stippled with goosebumps that raise the hairs around them like grasses in a breeze. My nose is cold and runs, and the tips of my fingers and cheeks feel the bite of the air. It is the bite of a lover, and I breathe white clouds into her as I scrunch along the shingly path and she folds around me.

In this distant cold I think of my children, who have chosen more complex lives than mine, and I worry for them briefly. This world is changing and will not be this way again. Am I the last of the simple ones? I tried to raise them to be free of hate and fear, of greed and envy, but they have chosen a world where those things are normal and I hope they have the tools to make their way through. Perhaps they see me as somebody who hides from that world, but at the end of my life I will be as happy and content as I am now. I do not see them often. The kids have grown and gone and are content; they come back when they're not and then leave again, feeling supported. That is how it should be. I am no Lear; their relationship with their partners is, and should be, far more important now than their relationship with me. My job, while it lasts, is to pick them up when they fall and send them off like swifts, back up to their lives.

I had been terrified of having children, because I didn't know what a father should be like, didn't know how to be a man – the only role models I had were characters in films, and people I didn't like. I was afraid to discipline my children; afraid I would ruin a delicate new life by allowing too much or not enough freedom; afraid even that I might be violent and aggressive, authoritarian and selfish, like my own father; I didn't even want the responsibility of acquiring a house to put them in. What I felt did not matter, and children came along anyway, and I did all those things and sometimes was too hard and sometimes too soft, but I figured it was never too late to fix things. I had wondered for decades, up until the day he died, if my estranged father would change his attitude towards me. All that wondering and willingness to forgive taught me that, even if I got it wrong, was too hot or too cold with my children, there would be years – decades – over which I could build a bridge between us, if I made the effort.

It still feels like winter. There is silence and a flash of sunshine breaks through the grey. I am waiting for the frosty days to pass, so that I can get on with the pruning before the spring arrives properly and the sap rises and the birds begin to nest. If I prune too early, then tender new growth may begin and frost will come and kill it; prune too late and I will disturb the birds that nest in the hedges and there will be no chicks and fewer birds and more greenfly. So each day I have to make a guess at whether there will be another frost. I live a sixteenth-century kind of life, but before I go to work I look at three different apps on my phone, and on Sundays I watch the farmers' weather forecast on the television to tell me what is coming. They often say different things, and I choose to believe the one that I feel is most likely. At the moment frost still seems possible – my skin and my nose tell me so. The window of opportunity between frost and nesting birds seems to be small to non-existent this year. I decide to leave cutting the hedges until the birds and nestlings have flown, and instead I prune the flowering and fruiting things.

Pruning will take up most of the next few weeks: apples and pears, raspberries and blackcurrants, roses, bush hydrangeas. Using a ladder, I am up in the bare branches of a big old Bramley, where the moss hangs thick and wet, taking off water shoots, and I rest for a while. Standing on a branch, holding another, I can see the sun glinting through the distant pines. Steam rises from the bright mossy cushions by my hand as the low, golden sun hits. The apple trees here are not healthy; they are very old and although they are cankered and split with rotting hearts, they produce a good crop of sweet and sour apples. Far too many to use. As often as there is bitterness in old things, there is just as often strength and sweetness too and, like the old roses, these trees are at their most productive. I dream of making eau-de-vie in an illegal still in the woods – Calvados that would warm me through the winter. I like strong spirits; a splash of whisky with my morning porridge keeps me warm in many ways.

I keep the trees neat and prune them perhaps every two or three years, mostly to take weight off the split and hollow branches to prevent them breaking. I was taught many years ago that you should be able to throw your hat through the middle of a well-pruned apple tree: the centre should be open to the sun and air, like a cup, a trophy, the petals of a tulip. I use what I learned years ago. You would think that, in gardening, nothing ever changes apart from the seasons, but knowledge changes and so practices change. We have learned in the past few years that trees are connected to each other through the mycelium that grows on their roots; that they share nutrients and information with each other on this underground network. Gardeners and farmers have always known that everything is connected, but people moved away from the land and forgot this. Now science is catching up with us and telling us the mechanism. I'm still a better weather forecaster than science, though.

Pruning Roses

Miss Cashmere is standing on her terrace, looking out. We wave and move away. There is cloud on the horizon – big white clouds with flat bottoms in a blue sky. The weather is getting noticeably warmer.

When you prune the roses, you just prune the roses. It is a silent occupation. Looking for a healthy bud that faces in the right direction, cutting off the branch above, at an angle, so the rain runs off the cut. Creating the best shape for now and visualising its growth, and the shape it will become as the bud grows. The hours pass, every branch is seen, every opportunity is offered to the plant to be its most perfect self. Not every opportunity is taken by the plant, of course. Unexpected frost comes and buds die back, but the thoughtful gardener tries to prepare for this by leaving an alternative opportunity – a bud lower down. I become the rose and feel where I might grow.

There are gardeners who do things by the book. I've met gardeners who prune their roses at the same time every year; they make a diary appointment to do it before this date or that weekend, and they feel stressed if the weather prevents them turning up. I do not do things by the book, I work by instinct. I would never last in a job working for the parks department, because everything is done by the calendar. I walk around the garden as I walk around my life, and if something needs doing that I am capable of doing, I do it. If it doesn't need doing, I leave it alone. Today I am pruning the roses because they need pruning; they are ready to start growing because the weather is right. I can see their tiny pink buds fattening, and although tomorrow would be good enough, I feel like doing it today.

To make the rose bloom, I manipulate nature by cutting it. I work lightly, first taking off any branches thinner than a pencil, then branches that grow in towards the centre of the plant, then I pull off any suckers growing from the base, and finally I look at the shape of the bush and make cuts high on the plant, above buds that are facing outwards. I want big, abundant shrubs covered in flowers, not the stunted, mean little sticks you find in parks and garden centres. I am not pruning for a couple of perfect show blooms, I am pruning for a mass of flowers that may not be worthy of prizes; who among us are worth prizes? A mean few pampered and cosseted specimens, some who, in the right place at the right time, manage to scrape through. I go lightly and aim for the joy of gorgeous abundance. Abandoned roses, as Vita Sackville-West says, 'throw their heads about as they are meant to'. Keep them three or four feet tall, allow them to be themselves, and they improve with age.

The rose bed contains about twenty pink and red bushes that I grew from cuttings. The skin of new plant stems is packed with buds and cells that are growing hungrily. Many plants make stems that are so vigorous that, if they touch the ground, they will start to take root.

79

A gardener takes advantage of this. In the summer, twelve years ago or more, I collected dozens of bright healthy stems about a foot long from some roses that used to be down where the summerhouse is. I took the leaves off, so that they did not draw water from the stems, and put the sticks, three at a time, in sandy soil in plant pots, on the greenhouse bench for the winter. By spring most of the stems had taken root, and I nurtured them into new rose plants in a nursery bed by the greenhouses. As the collection grew strong, I made this new rose bed and transplanted them, and they have been there ever since, growing older and stronger.

I top the soil with living compost every year, and by the end of autumn the rose branches are thick and tangled and covered in rosehips. Unlike other gardeners, I abandon them in October to leave the fruits on over winter for the birds and the squirrels, and for the joy of seeing the bright-red berry against the snow.

A morning spent pruning roses makes my hands sore. My right hand hurts from using the secateurs on thick, woody stems and is covered with cuts and scratches and spots of blood, the first flash of red from these thorny stems. I hold the branches that I am removing in my gloved left hand, but my right is uncovered so that I can feel what I am doing. A glove would take away the sensitivity of touch.

Hand, senses, shears and shrub slowly work together to find the happy shape. I am alone in the frost – me and this quiet world, just looking at each other slowly, pausing and snipping. I'm a child again playing in nature, under the sun, standing on frosty grass with the blackbird, hearing the woodpecker and crow and rook echoing across the empty, cold fields. On this still late-winter day the sound travels in strange ways: a distant train clatters, the nearest track is four or more miles away, and the cold air holds its rattle close to the ground.

Snow

A flat white moon stares at me through the window. I focus hard and can see it moving as I imagine myself floating in a pool. Breathing out, I can feel myself sink; and as I breathe in, I rise and float and I sail through the night, with Peggy breathing softly next to me, and I don't feel alone. As I approach the inevitable shore of morning, the first birds begin to sing and there is a strange light in the sky. The slatted light comes through the blinds and lays stripes on the white bedroom walls and duvet. The colour is not the usual yellow-white, or grey, it is a colder blue-white. The morning is quiet, the sounds dulled. I scramble out of bed to check and pull the slats of the blind apart, to see that the world is covered in thick snow that is still falling and is deep enough to last for days. The sky is light and the roads are silent, and I listen to the muffled bird-song and her soft breathing, while I wait for Peggy to be washed up into the day.

Today no gardening can be done. Later we paint the walls and stack the bookshelves that I built for her last weekend, and I am enjoying the fuss and the fluster as she plans what books should go where, and how she chooses. I don't understand why she organises them in the way she does, but it makes her happy to have the books in a particular order: novels here (author's surname first), reference here, books to read here, books she loves here;

and I wonder what that says about the other novels that she keeps and doesn't love and are not used for reference. I prefer my books chaotic; we have separate bookshelves and I get rid of books I don't love. I want to be distracted by something I hadn't thought of, when I am looking for something. I crave random encounters in my ordered, seasonal life, but only as far as books and nature are concerned; random encounters with people are something I try to avoid: I am clumsy and often misunderstand the purpose of talking with others.

We are painting our room and it feels like spring. We're enjoying our freedom, and we are happy. We spend the rest of the day reading and writing, then crushing through the snow to the pub in the village. Everyday things – this chatting and going for a stroll, washing up, watching the sun set as night comes in – are the best part of our lives. These ordinary days are delightful; they are what our lives are made of, like arches on a bridge.

Peonies

It has been sunny; the wind has dried the surface of the snow to ice, and beneath it the earth has been protected from the drying air. The sun's rays have filtered through and melted the snow from the underside, forming an insulating crust above. Scabs of ice on the grass and the flower beds have become tiny greenhouses that have warmed the soil below, and under the dripping, melting snow things are beginning to grow. When I stand on the few remaining patches of ice, they collapse into the hollow air beneath, where the grass is fresh and green. In the flower beds the blood-red shoots of peonies (*Paeonia*) are coming through; strong, closed baby-hands reaching for the sun, bunches of green delphinium (*Delphinium*) shoots with crinkled, hairy leaves push through the leaf mould from the crowns left in over the winter. The daffodils (*Narcissus*) are full and happy, and snowdrops and fritillaries (*Fritillaria*) are flowering at the base of the horse-chestnut tree in the lawn.

Buds on the bush hydrangeas (*Hydrangea macrophylla*) are fattening and it is time to take off the dried flower heads that I left over winter. Traditionally gardeners leave the dried flower heads on to protect the new growth from frost. I leave them on because I prefer the look of shady abundance to the stalky look of a pruned plant.

That is why I also prune my roses later than many people. I don't like the look of cut ends, and prefer to cut the stalks just before they start to grow again, so that they can't be seen, to maintain the illusion that this is all natural. The dried hydrangea heads, and the poppy heads that grew in the soil below them, are slowly turning to dust and provide a home for earwigs. Earwigs eat new and soft plant growth and love to munch on dahlias, but they also eat aphids. I leave the earwigs in their crumbling poppy homes; they'll fall to dust soon enough and, as new growth comes, I'll cut down the stalks. But I'll take the hydrangea heads now, so the new buds have air and light around them.

I am wrapped in layers of wool, as the air is still and cold. My breath makes clouds. I am wearing gloves and, throwing the massive dried hydrangea heads into my green barrow, I remember William Carlos Williams's perfect poem 'The Red Wheelbarrow', on which so much depends – this simple machine, a lever and a wheel, allows me to move heavy quantities of stuff around. I wheel them to the compost and start a new heap, then back to deadhead the daffodils. I take the heads to the heap, past golden bunches of small willow catkins and long, drooping hazel catkins and massed excited birds, and pink daphne flowers clustered on bare branches. By the house the scarlet flowers of the Japanese quince (*Chaenomeles japonica*) strike me, and forsythia (*Forsythia x intermedia*) is blooming yellow along bare stalks. Spring is really nearly here.

Next week I will sow sweet peas (*Lathyrus odoratus*) outside. Two at a time, under a pyramid built from hazel or willow sticks and green twine. I'll soak the seeds overnight, then pick off the weaker plants as they grow, so that the strong can survive.

Potatoes Rattle in a Pan

The rains come, the last few patches of snow that had been blown into the dark places where sunlight never penetrates melt and are washed back into the earth and, because of the weight of the rain, I pack up and leave. The cold rain becomes torrential on the windscreen and roof of the van; the sky is black as I drive home through country lanes that are old and have drainage systems that were meant for dirt tracks and farmland, not tarmac roads and houses with paved driveways. Flood-brown soil runs down the road, deep, and I create a muddy wake as I drive. As I wait at traffic lights nearly home, the light – red, then green against the dark sky – smears on the glass and flashes brightly through the waterfall running down my windscreen as the downpour pounds on the roof.

I am a working man with broken nails and skin-cracked fingers, and at home on my chair I scrape off the mud and brush wax into my boots. I can grow new skin; I can't grow new boots. I polish them as I polish my shoes that I go to the pub in, on Saturday night. Polish them to a high gloss, as my father, who at one time had been a soldier, taught me, and as his own soldier father before him had done. All dead. All history. Looking into the shine, I see a vague reflection. The features could easily be those of my father or my grandfather, smudged and warped by the curved toecap and brown wax.

I love these quiet days when the rain splatters on the wet flagstones like typewriter keys and makes little ripples an inch across – days when I can throw open the doors to watch and feel the cold air rush in, to chill my bare feet. Peggy is upstairs, moving about, and I'm glad of her far-away company and the way she leaves me be. To just be and watch. When we met I was a wild thing, barely socialised, and she was a wild thing too, having grown up wanting to run away from the mountain in Wales and the lonely cottage in a field near a lighthouse where she lived and played and read fairy stories alone. We vibrated with a similar frequency, so we moved closer together like magnets, and made children. They came into the world and they were like both of us, but were new and like neither of us, strangers who we came to know and love.

Over the rest of the day the sky lightens and clouds roil in the fierce wind that comes and blows the rain almost horizontally. The bluster continues through the night; things clatter to the floor outside, and plastic things blow down the street: dustbins, garden chairs, rubbish bags. The living things stay put. Something bangs rhythmically somewhere – a shed door perhaps. Peggy cannot sleep. The rain continues. In the morning a new river runs down each side of the street, boiling out of the grid-lids in the gutter, and the wind continues to howl. The electric lights stay on all day, the heating is roaring, as raindrops the size of marbles bounce off the shiny black road, and Peggy and I remain indoors, reading and enjoying the rain, and we steam up the windows with our cooking.

Mimi, my cat, curls next to me; she licks herself elegantly and twice my hand, unbothered. I watch the rain and listen to the distant traffic pass, contentedly doing nothing, calm and sinking slowly into my sofa, bare feet on a Welsh tartan blanket, brown and green, a sett no clansman ever wore. My body aches from labour and a poor night's sleep, and I drift down into the sagging cushion, like sand. Flesh against fabric. My coffee slowly cooling as I fail to make the effort to reach out for the mug – the red mug my daughter bought me. The cat moves a little, starts to snore, twitches. The distant traffic's quiet now, arrived at its destinations; the vehicles are cooling on their home parking spaces, empty but for the bits that people shed and leave behind: a coat, a bag, an empty crisp packet, crumbs. My coffee's cold and still I haven't moved. I can't – the cat is now on my lap. Peggy brings a fresh cup. Potatoes rattle in a pan. Peggy talks and giggles on the phone in another room.

Cherry Buds Appear

The slightly warmer weather has brought a dewy, misty morning, the earth still wet from rain and meltwater. The buds on the cherry trees are starting to show a little colour, but will not open yet for some time; the brambles and the roses are sending out new bright shoots; the red-tipped tulip leaves, all clustered together in their bed, are about four inches long, and in a couple of them I can see the small flower bud and stem tucked inside the leaves. The daffodils are still strong, and I wonder at how they managed to stay upright after the storm, which lasted for three days. I spend the morning deadheading them.

Wet plants, wet grass, wet boots. The crows are calling from the treetops; wood pigeons and sparrows sit damp and silent in branches unmoving in the still air. It is one of those quiet mornings when the grey air seems like thick felt, trapping and dulling the sounds of the world. I can hear my trudging steps; the crisp snap of the daffodil stems as I break off the seed pods; my breathing. Over the hills there seems to be faint pinkness in the high-clouded sky, the kind of pinkness that sometimes means frost, but there is no frost. The blanket of thick cloud keeps what little warmth there is close to the earth, just enough to prevent ice forming, and I snuggle down in its softness to a day's restful work.

I feel sun on my back as the cloud breaks. A blackbird sings. The plum blossom is open pink and white, and hangs in weighty clouds from the ends of knotty stems that spring from cracked branches with splitting bark on timeless trees with the look of centuries – a fixed broken state of perfection. The leaves of the crab apples are about to open, their folded tips bright red against the grey of the sky and the darkness of the bark. As I sail through this ageing and new blooming, I feel that it is good, so very good to be cracked and mossy and calm and solid in the centre of this maelstrom of nature, swirling by day, by hour and by minute. I have the time to watch it now. Youth has no time; it is rushing headlong into its own future, but I can put the brakes on and watch the world slip by. The better I watch it, the slower it goes. I can walk and feel and sniff and taste the air, the sun, the frost as it flows across my skin and into me and out.

I am like a child catching moths. The moment is flighty and lands in my open palm and immediately flits away to make space for another, each as generous and kind as the last; a million, a billion moths that each time leave as soon as I have seen them.

The equinox will be here in a day or so, when day and night are of equal duration. I lock the gates and leave for home as the light begins to fade. I get in through the door and it is still just light enough to read. The house is full of her: in the hallway her umbrellas; and on the sideboard that she bought sits an old wooden bowl she got for me second-hand, where I put my keys; a painting of her in the kitchen that somebody did when she was at art school; her warmth and scent, and the steam coming from the stove, candles lit on the table; her coat over a chair, and her glowing and waving a spoon around.

'Where have you been? It's getting late.'

'I didn't notice the time,' I said, 'the days are getting longer and there's lots to do.'

'I've been thinking about our boy. I'm a bit worried – we haven't heard from him in a while.'

'I am sure he is fine; he's busy, that's all.'

'Yes, I suppose so, why don't you ring him?'

'I'm tired,' I say, sniffing around to see what might be cooking. 'You ring him if you are worried.'

'Yes,' she says, 'will you keep an eye on the dinner?'

'Sure,' and I wander over to the stove, where there is a pot of soup bubbling away. We eat a lot of soup.

I remember him, my cherry bud, being born, greasy and bloody and looking into the lights as soon as he was pulled out; and her, my lovely wife, torn and tear-stained and exhausted. And I remember crying for the agony of it all, for her agony, going outside so she wouldn't see. He is a man now, but still looks into the light, and she still hurts from time to time. Kids will hurt you, they will take all your love, then all your money, and all your time and attention and your work, and will ruin your stuff; then they will leave and break your heart, and you never stop being afraid for them or loving them, even when you hate them and they are strangers.

She calls him from another room and in a moment she is laughing and happy, and back with news that his business is going well, and he and his girlfriend have been driving to the south on his weekend off to find some sun, and he would like to go and live in Italy.

The Middle Way

Equinox. The first day of spring. Everything changes in the garden and nothing changes. The blackbirds and sparrows are going wild, and there is a light shower, so I stay in bed for a while until it passes and I read. I am reading poets who rage: Charles Bukowski and Dylan Thomas. I have no rage. I refuse to lose my love. They seem to have a certainty that I do not, they have clarity about right and wrong. I do not trust such inflexibility; the truth is always deeply buried in the middle, where it wanders about, vague and unsure of itself. The poet's anger sparks my own for a while, and I watch it rise and see my leaning in that direction, but I don't water those seeds of discontent; I let them wither and die, and the morning passes. I love these ranting poets with their wild passions, who surf on their emotions and drink themselves to death, but I choose to keep my distance. I know so many men who have angered and drunk themselves to death. Unhappy souls, who are generous with their misery.

Each one of us is a mix of strong and weak, good and bad, dark and light, and when they get out of balance, the whole of the person is out of balance and can easily fall. I feel out of balance after reading these poets, but I do it anyway. I'm on a tightrope and I like to make it sway sometimes for fun. Then, as I absorb their feelings and I have had enough, I correct the leaning, quieten my mind and make more room inside. It is like pouring stagnant water from a jug, so that it can be filled again with fresh. Washing the dishes after a meal, tidying up.

I was brought up to believe that I was worthless, and nothing I did was of any value. It was a daily lesson. People who are brought up with such beliefs – that they are stupid, ugly, weak or fat, or even clever, strong or handsome, athletic or hard-working – all have a lean in the direction they were taught. That lean is a delicate and addictive thing, and sometimes there is nothing to stop the leaning become a heartbreaking, dark, slow fall into self-destruction.

It is dangerous to have fixed ideas about yourself: call yourself a writer and have no success in writing; call yourself a steelworker and lose your job; call yourself a waiter and what are you, when you are not waiting? Better to play at being a steelworker, play at being a firefighter, play at being an artist – take on the role, then shed it when you choose. When I am in the garden I am the gardener, I do the gardener's job, do everything the gardener should do, but really I am playing at being a gardener. Today I am playing at being a writer, a lover, a poet, but in essence I don't feel myself to be anything different from anything else that exists in the world; just an emanation playing. Human beings are at their very best when they play, and here in this garden my work is play. I choose it that way. I know that nothing's real or fixed. Playing 'let's pretend' that I'm something other and separate from the flow. I play with soil and seed and flowers, with trowels and forks and spades and mowers. I play for the joy of playing, as small children play before they are taught to compete. An emanation playing at this, then this; and when I am not playing, I am simply a potential ready to play: a child, a wind, an empty jug ready to be filled. In the end, all we are is our attention, there is nothing else.

There is no competition in this play. If I am struggling or in conflict with the way things are flowing, then I am doing it wrong and need to change direction, find a flow that is going in a direction that is preferable. Whether I like it or not, I am like the hare and the owl, the carp and the buzzard. The creative universe sings its songs, and that is all there is. Filling life with other stuff does not change that – it only hides the truth. Life is full enough; it is plenty for me to watch a cloud grow or pass, a speeding flock of swallows, a swirling mass of starlings, the waves on a beach, a blackbird sing, the people pass.

The plants and the trees, the bees and the sun and the rain are all helping me, and at about eleven o'clock the rain stops and I head for work. While driving to Miss Cashmere's, I think of the men I knew who drank themselves to death, who died or were crushed. There are many ways for a man to die: working men hollowed out by hope and poverty; intelligent men denied education, who found freedom in alcohol or drugs or suicide. David, who was twenty-four when he hanged himself from copper wire attached to the banister in his shared house; he hung, dangling and stretched for his housemate, an art student, to find, who painted the scene over and over for as long as

I knew him. The dozens of men who turned to drink or drugs and who, on that day of turning, stalled their development as human beings and, for some, began their end – they were not playing.

In my school and at home, questions were seen as signs of dissent against the hierarchy. Only the innocent asked questions; school, family and work taught us that asking questions of our elders and betters was rude, impolite and would be punished. As we grew older, many young men abandoned questions, but a few I knew thought about things too much. Some managed to give themselves an education by joining the steelworkers' union, reading Karl Marx or George Orwell: *The Road to Wigan Pier, Down and Out in Paris and London*, then *Animal Farm*. They realised their subjugation, became political, but stayed in the hierarchy, ready to fight, but this time with the class system. Others, who didn't read, became filled with burning white hatred for anyone who was not like them or who looked at them funny. All those fighters were people whose role in life was to keep the wheels turning until there was a war, and then to go and fight in the war. They were educated and bred to be easy to manipulate, by a government that would use the media it owned to flick the switch on their emotions and lead them to wherever they wanted them to be. Men rarely want freedom from the herd; it takes more courage not to follow than the herd can ever imagine. You cannot tell a man to fight, but you can make him want to fight.

I never wanted to fight. I remembered my father watching the boxing on the television and, seeing the bow-tied, cigar-smoking men in the audience, I was struck with the idea that when men fight, other men – men who are not fighting – are the ones to benefit. Some of the men I grew up with stayed angry, ended up dead or in prison, or hurting people, never able to find peace. We were all deliberately sown with seeds of fear and hatred, but I choose not to water mine. I leave those seeds in arid ground: the racist, xenophobic, sexist, homophobic beliefs that I grew up surrounded by. I will not give them my attention, will not allow them to take root in me. I'm always learning to spot them and send them back to the depths, when they show themselves, which they do from time to time.

I am saddened for, and fearful of, those people who do choose violence and hate; they are everywhere, and each night before I go to bed I check the locks.

Today the day is equal in duration to the night. Tomorrow there will be more light than dark, and things in the garden and air and soil will grow faster. Tonight we will celebrate with whisky and wine and food and fire. It looks like Miss Cashmere is celebrating, too. There are bunches of daffodils in vases on the window ledge, and she smiles and waves at me from the window, then struggles to her feet and heads to the kitchen door for a chat.

There's a doll-shine to her wrinkled ski-slope skin, and again I'd rather not talk. Talk is not my chosen medium. I am unskilled at it. But as it seems I have no choice, I switch off my concerns and trust to instinct to say the right things.

And all we chat about is 'How are you?' and 'Isn't it lovely?' and I ask her what she's up to today, and she's having lunch with a friend in town; there's a taxi coming soon and she's ready and brave in her dark-blue knee-length dress and green coat, and her hair in its bun. And we talk about nothing at all, and it feels good, and I enjoy it, and realise that words were not what passed between us, but love.

Sparrows Begin
to Nest

The wood pigeon has found a mate. I can hear two of them now – the one that has been here all year, and today another one with a much deeper and slower voice has joined in. The blackbirds have been singing all day since the day before yesterday. They were singing before dawn this morning, and late into last night. I threw peanuts out for the rooks that live in the trees opposite and are learning to recognise me. One of them came down straight away, and the other stayed on the ground further away. The buds on the cherry trees grow fatter every day.

Two or three hundred daffodils remain, and Miss Cashmere comes out with her newspaper and sees me standing looking at them. A handful of them are blind; they have no flower stem or bud, but are just healthy strappy leaves, and I make a mental note to put some fertiliser down to feed the bulbs for next year: blood, fish and bone, which has a smell that only gardeners know and love. A few years ago – I can't remember how long, five or six perhaps – I bought five sacks of Welsh daffodils and planted them in great drifts: 450 bulbs in each sack. Casting them by the handful in brushstrokes, then kneeling on the earth and digging a hole where each bulb landed; all to the same depth, so they would come at once, placing each bulb into its own private nest and covering it with earth and the circular divot of grass.

We stand and look at the field of yellow heads.

When I was a vagrant, wandering like a friar with a pack, and having nothing to read but the weather and the landscape, I came to understand that the earth is a library: stones, trees, animals, scents, water and winds are some of its books. Each sound, each temperature, each sensation on the skin or underfoot is part of the story. If I had my eyes closed and was dumped here from a time machine, I would know it was early spring. The air is cold, there is still snow on the ground in the darkest sheltered places, but there is something else – a scent maybe, the light is different, the birds are noisy, the sun crawls higher, the day grows longer, the air smells sweeter. Tiny sensations. I let my internal dialogue go quiet and the world continues to chatter, and I listen to those stories that the world sang me to sleep with, as a tramp.

It is easy to distinguish between the scent of a lake, the scent of a fast river or a slow one and the scent of the sea. Easy to work out in which direction they are, how big they are – once you tune into them. All you have to do is be quiet. You don't have to try and learn; they teach you, and the knowledge creeps in if you just listen and notice. Bodies of water change the colour of the light in the sky. The colour of the air above towns can be seen from dozens of miles away, even in daylight. The birds tell me that there is a cat nearby or a human; their alarm calls for a man are shorter, but when there is a cat nearby they seem to call louder and for longer. I do not need to know these things, they are simply the story the landscape tells, part of the constant chatter of the world. The land is constantly moving, and it tells you its past as the lines on a person's face tell you of their life. A drover's road fills with water and becomes a string of small ponds, alive with frogs and newts and bugs. I have no idea how these creatures find their way there, but they do. An abandoned railway line becomes a strip of woodland where bats and badgers find safety. An hour or so resting and looking at the things around tells us a story of an ancient living place that never stops moving, whether we care for it or not. Whether we are here or not. Nature loves to change the things that she has made and make new things from them, remaking all the things in the world and spinning them into something else.

Later, when I wandered into town, I realised that people were books, too. They are the roses and peonies, thistles and rocks, colour and breeze. Like petals on a spring wind, scattered here and there. They do not need to speak; it is in their faces, in their posture and movements, in their clothes and shoes, and in their skin and its wrinkles. Today, as we look at the daffodils, I can see that Miss Cashmere's hair is loose and not as tight as she usually likes it. Her skin is pale, grey even. She is not smoking. She smells different. She usually smells of tobacco and soap, and some kind of powdery cosmetic smell that I do not know. Lily-of-the-valley perhaps.

I like smells: a rotting plant, fresh and old compost, manure, the pissy smell of a runny French cheese, the woody, smoky, iodine or fruity smells in different whiskies. Today she smells musty. As if she has not been out in a while; she needs airing. There is darkness around her, too. She simply says, 'Morning, Marc' in a subdued way that has none of her usual chirp to it.

'Morning, Dorothy,' I say, matching her mood.

There is little response, a short pale smile, her lips white, her skin loose.

'The daffodils do look lovely, don't they?' I say.

She looks older than she usually does, and she stands by me and looks at the daffodils. 'You do a lovely job, Marc,' she says as she wanders off to the summerhouse with her newspaper.

Bees

Masses of small bees spinning off and away from the hanging pink flowers of flowering currant swinging in the breeze, and flies around the compost sunning themselves on the upturned barrow. In the twisted thicket of rambling roses the blue tits and great tits and sparrows are sorting themselves out and singing, safe from the cat (eyes too close together), who slinks as a fox across the shed roof. Spring has arrived and it is warm.

The sun falls and the cat, frustrated, howls – waaouw, waaouw, waaouw – at bats chasing insects, circling at dusk in the rising hot air, feeding on masses of them. Twilight lets the shadows out and transforms the mundane into the magical; the everyday is in balance with the unknown, and the sacred in the banal becomes more visible. The barrier between fact and fantasy fades and blurs. This is my natural habitat and, when it isn't there, I tap away at my keyboard to create it. I love the blur when two worlds merge, two people connect, two languages try to communicate, seeds meet earth and life begins.

Daffodils

You're right
The daffodils are beautiful this year and he would have loved
    them
And he is not coming back
Because he can't

Words can't help you
I can't help you
It is not a new beginning
And life does not go on

A stalk of grass remains
The mower missed
I snap it off as I pass.

Narcissus – Are You There?

We will lose those that we love, and that will hurt. We will die and, for many, dying will be physically painful. All this will end, and we will want things to be different. That is as it should be. Everything passes and this perfect moment will not be repeated in all the history of time. Nor this one.

It is sunny and warm and she sits on a bench at the top of the garden, supposedly reading her paper and smoking, but I can see from the corner of my eye that she is watching. Flawed beauty, impermanence, incompleteness and suffering. She is alone and spring has come. She is in a long warm dress, a woolly cardigan, and there is something about the way that she looks down and across the acres, wistful and dreamy, holding her cigarette, and I think briefly, curiously, that perhaps I am a narcissus and I see myself in her isolation. There was a brief sense of fullness and tenderness at seeing her, and I imagine the garden without her and I am bereft. I want to see her wander across the grass again and again. I want to see her go about her life. She completes this place for me.

I go bending through the daffodils, deadheading them, pinching off the ovaries so that they cannot make seed and instead put their energies into making the bulb bigger and fatter for next year. I throw the browning yellow heads into a big green plastic bucket and, as always, am surprised at how many I collect. I need a bigger bucket – too many trips to the compost heap. Nipping the heads off with finger and thumbnail, so that at the end of the job my thumb is green and layered with a crust of dry sap and the drying scent of daffodil. They have passed their best and many of the heads are turning brown now, as the tulips and the clumps of grape hyacinths (*Muscari armeniacum*) come through.

The daffodil is properly known as the narcissus, named after the myth of a Greek hunter, a man known for his physical beauty and his obsession with it. He had a string of admirers, whom he rejected as not beautiful enough. One of them, Echo, a mountain nymph, followed him around; and Narcissus, feeling that he was being followed, shouted, 'Who's there? Who's there?' Playfully, flirtatiously she repeated his cries: 'Who's there, who's there, who's there.' She was not beautiful enough for Narcissus and he spurned her. Desolate, she faded away and became nothing more than the echo repeating in mountain glens. In another version of the story, a young man called Ameinias fell in love with Narcissus and he, also rejected as not good-looking enough, took his own life. In both versions Nemesis, the goddess of retribution, saw how cruelly Narcissus treated his suitors and led him, thirsty, to a mountain stream; there, leaning down to drink, he fell in love with his reflection and remained there, looking into his own handsome face until at last he stopped living, stopped breathing and, taking root, became a flower – a daffodil, forever looking down.

'Narcissus' comes from a Greek word meaning 'numbness' or 'narcosis'. The plant is toxic, the bulb most of all; it contains a poison called lycorine that will give you nausea, vomiting, diarrhoea and stomach pain for a few hours. Narcissus bulbs look a little like onions and have been eaten accidentally, and purposely, in times of famine, by starving humans. They also contain chemicals called oxalates, which are microscopic crystals shaped like needles that cause a burning of the lips and throat. Many foods contain oxalates, and these can give you kidney stones – spinach and rhubarb, for instance – but if you combine them with dairy products, they don't do you any harm. Rhubarb is healed by custard, spinach is cured by feta cheese.

She is still there, watching or daydreaming as I finish the daffodils and head to the shed where I store the lawnmower. The story of Narcissus cycles around in a loop in my mind and I try to understand it. I think about its obvious meaning: how self-love is destructive and creates a feeling of numbness towards other people, and of the poisonous daffodil bulb that causes numbness; but also about the delicacy of our emotions: how the feeling of love can make us tender and enable us to communicate with touch in the most delightful ways, and how the same feeling can control and totally destroy our lives and those of others. The dark side and the light. We all have a dark side and a light, and the only way we can feel whole is to love them both. I have known people damaged beyond all redemption by the corrosive power of self-love, who see no darkness about themselves, and others also damaged who see no lightness. Balance and harmony are everything.

Minotaur

I pull the old green lawnmower from the shed where it is stored over winter. A big, heavy petrol-driven machine with a roller that makes stripes in the lawn. I had serviced it, as I always do, before I put it away; emptied the petrol tank, changed the oil, sharpened and lubricated the blades: six sharp cutting edges coiled around a cylinder that spins and slices against a fixed lower blade. I set the cutting height and fill up the fuel tank and, using the muscles of my back, pull the frayed cord that turns the engine over. The heavy flywheel turns, the piston rises and falls, a puff of smoke comes from the exhaust and the shafts and gears come to rest again; another pull, spinning the mechanics around and sending fuel into the cylinder and oil round the bearings, another puff or two of smoke and another resting of the machine. On the third pull, as it always does and always has done, since we bought it new about fifteen years ago – she paid, I chose – the spark ignites the fuel and the motor starts to spin on its own, and the simple single-cylinder, side-valve engine, built for reliability rather than performance, chugs away happily in its familiar beat. Ease it into gear, drive it into position, move the lever to engage the cutting blades and off we go for our regular long, slow waltz together. I have travelled this same circling path behind this same mower every week during the growing season for fifteen years. Before that, there was another green mower that was ancient and marked and scarred by the lives of past gardeners, when I met it, and that one died in my hands.

The late snow has pushed the mowing season back a couple of weeks, but now the grass is good and strong and ready for its first light cut. Once the average temperature rises to eight degrees the grass starts to grow; if a chill comes, it will stop growing again for a while. At first it needs cutting once a fortnight, then a regular weekly mowing – if not twice a week, when there has been sun and rain and the grass has grown faster.

The noise of the mower disturbs her, takes her out of her reverie and she uncranks herself, struggles upright and slowly goes inside. Sits at her dining table facing the large window, looks out at the garden, watching the entertainment of this familiar man go backwards and forwards behind this bright, noisy machine. She sees that I see her watching me, yet she continues watching and I feel exposed and look away. I feel naked and want to conceal myself from her gaze with an animal skin. Her cat sits on the wall and, unselfconsciously, he watches me too for a while until, spotting her other cat, he fixes his gaze there and they stare at each other intently.

Mowing the bright grass explodes a scent-bomb that wraps a green blanket around me for a while, then fades as the nose gets used to it and the oily green scent of grass and mower become part of the background – a smell I carry home on my clothes and skin that Peggy will notice and enjoy. The heavy roller behind the cutting blades

III

makes stripes on the immense lawn, bending the grass, one way as we go up, and then the other way as we return back down. Up and down in parallel lines, then around when we reach the wall surrounding the pond, like brushing velvet, or a mole's back. Each time we mow this lawn we travel eight miles. The mower goes slowly, I follow slowly behind it, stopping every four stripes to empty the grass collector into the trailer, which I hooked to the back of the little tractor that I parked on the lawn nearby. When the trailer is full, I drive it down to the compost heap and tip it on top of the deadheads from the daffodils and the hydrangeas.

After the first few miles I begin to imagine myself a Minotaur in a labyrinth following a path. The Minotaur, said to be trapped in a maze by the gods, is the offspring of a queen and a handsome white bull. He is half-man, half-beast and devours young men and women for sustenance. He is the unsophisticated creature at the centre of us all, and the journey to finding him is a spiritual one. For a human to follow a labyrinth is said to be a meditation, an exploration of what it is to be human, a pilgrimage. I wonder for a while if the pilgrim is the monster or the sacrifice, before realising that the whole point of the story is that he is both. In order to follow the mundane labour of the looping path that offers no choices, the walker has to sacrifice himself and become a beast. A few miles further on, I lose myself. She watches me become a beast, from her elevated position, but I have long since forgotten she is there, forgotten everything. I stop to eat and see her watching.

As green as it should be, the mower ticks as the engine cools. Hot vapours of petroleum and grass while I eat my apple. A pool of sunlight prints the crinkled edge of the acer on the mowed stripes. Unfixed, it moves and is gone. A squirrel runs across the path and I go back to work.

An occasional bee wanders over looking for clover – he has no path. A man used to come a few times a year to spray chemicals on the lawn to make it green, and kill the moss and the dandelions and nettles and trefoils and buttercups and chickweeds, the cleavers and clovers and docks and daisies, and all the million and one wild plants that grow in lawns. When I first started here I told Mr Cashmere that I wouldn't use chemicals, I didn't want to poison the earth. He said, 'Fine, I will get a man in to do it', as if I were not a man. It was his garden. His bit of nature. His poison.

There are chemicals available to spray lawns with, so that it doesn't grow so quickly; others to kill the worms and beetles so there are no worm casts, no moles feeding on them. There is a company that will come and spray your lawn with a pigment to make it look green in the summer when everybody else's goes brown; others that will dig it all up and replace it with plastic grass, which never changes colour and smells of vinyl in the hot sun, and will stay looking the same shiny plastic green for all of eternity. The glossy full-colour leaflets for these poisons arrive, uninvited, through my letter box every spring. These are for the people who are not gardeners, people who want to control nature.

To speak of controlling nature is like the waves wanting to control the sea, the song singing the thrush, the flower creating the earth. We are not the sea, we are not the thrush, we are not the earth. We are the wave, the song, the flower.

There are about a hundred different grass species that are seeded in British lawns, and nearly 12,000 species of grass in the world. Few people care, as long as it is green.

The early spring sun browns the back of my neck as I mow, and I take off my heavy tweed jacket and hang it on a crab-apple tree at the edge of the lawn and work in my blue workman's shirt with the sleeves rolled up. When Miss Cashmere was away on holiday once, I mowed round and round, starting my journey at the pond and working outwards in a spiral that became waves at the edges of the irregularly shaped garden. The patterns had been rubbed out by new growth and wind by the time she returned. The stripes only last a few days before they need making again. I could, if I had time, mow it again crosswise and make a chessboard pattern, a green tartan, but I haven't the energy and there is too much other stuff to do, so up and down it is. There is no way to get lost on this journey, no choice to make, just the path from entrance to exit. The seasons come around, turning the corner at the solstices and equinoxes and spiral round again – everything turns around. Every couple of weeks a spinning blade in an uncaring machine slices off the top of the grass. The grass never gives up, it simply keeps coming; it never asks about the point. I have become like the grass. I keep coming, I don't ask about the point.

Depending on how long the grass is, it takes me between five and six hours to cut and clear the cuttings to the heap where the grass will start to decay and go yellow within half a sunny day. As it decays, it warms up and the middle of the heap becomes hot. The heap is constructed in layers, so that the heat of the rotting grass kills off any weed seeds and breaks down the flower heads and cuttings that are placed on the other layers. I cover it with cardboard and tarpaulin to keep the heat in for as long as possible, to let it cook nice and slow all summer long. Slow-worms and grass snakes sometimes nest there, skidding out of the way and fading into the grass like memories as soon as I lift the cover.

Grass cut, mower parked by the shed to cool off, I take the edging shears and sidestep around the island beds, trimming the long grass that hangs over the sharp edges between lawn and soil, and let the grass fall onto the beds, where it will turn yellow. Some will be drawn into the soil by worms whose naked bodies go through the soil, while the soil goes through them and comes out changed, and the rest of the clippings I'll push in with a hoe.

As the day ends, I find a small bay tree that has seeded itself in one of the flower beds and go to get a spade so that I can dig it up, but on my way I change my mind and decide that I might leave the little sapling there and train it into a lollipop, and so I go back to the tree to inspect it further to see if it is suitable. It is in a good place and of a decent shape. I tidy it up a little, cut off some lower branches with the secateurs that always hang on my belt, shorten some others so that it will thicken up.

Thin white clouds are splashed across the clear blue, oh-so-blue sky, carelessly perfect, done by a master with a Japanese brush; a crow caws in the deep silence. I return to the shed with my tools in the twilight, moving through flitting and stationary, silent insects: hoverflies and bees; my feet make the only sounds, the hand-fork and trowel clank together just once as I put them in the bag. The birdcage-house is lit up. I turn the key and start the motors and make an unwelcome noise as I drive the mower and tractor into the shed, and then silence again that lasts for miles. Pack away my secateurs, put on my old tweed jacket and sling my canvas lunch bag over my shoulder. You can see me now, in your mind's eye, you know exactly what I look like. Beard, tweed, earth, blue sky.

April

Distant Thunder

I am drinking black tea with a spoonful of jam in it, a habit I picked up from some Russians I met. Last night I drank vodka, lots of vodka, and Peggy is away for a few days, so I got up late and I am rehydrating and thinking about going for a hike in the mountains. I'm sitting by the open back door, reading and planning what to make for dinner when Peggy gets back tonight. I send her a message on my phone asking if she would prefer something hot and spicy or salad and cheese and pickles, and I feel disgusted at my constant reaching for technology and want to get away from it – away from the stuff and clutter – and meander in the hills and pitch up for a night or two in the open. There is a fear and a frustration, and although the garden is busy and I am already tired, I start to think about making a decent solo hike when the weather gets better, maybe even for just a night. I wonder if I will make it happen, and already know I won't. I have never felt like this before.

I don't usually have much call for concern about things, for my life is simple and inexpensive, but now as I have grown older, I find myself spending time organising myself so that I don't lose my keys and glasses. I discover myself making plans, but I have rarely in my life made plans. I used to think that one either made plans to do a thing or one did the thing. I felt that making plans was a way of not doing the thing, but pretending that you had. I used to change my position, move from one place to another and view the world as I wandered; but more and more I find that my position remains the same and I view the world as it goes past. I do not wander any more. Those words come and scare the living daylights out of me: 'I do not wander any more.' In years gone by, I would pick up my jacket on the way out and not return for three days. Now I make plans for things that I do not do.

The church bells down in the valley, where I can still see the steeple poking out between the trees that haven't got their leaves yet, ring three times, then a pause and then another three. Maybe there were three more, but the sound got blown away in the wind that bends the bare trees and howls across the top of the chimney, and raises the ash in the cold, empty stove. Cars arrive and depart in the street outside my door; and voices, as people turn up, or leave to have Sunday lunch with family or friends. At least I am alone. The wind and seagulls scream in seagullish abandon as they look for mates and offer potential partners gifts of lumps of grass; they also give me the idea that I could go to the sea instead of the mountains, sit on a rock and watch its emptiness. I'm craving connection, it would seem, and so I make beetroot soup and sour some cream, and the cooking becomes a meditation and the connection to the moment comes.

The food is made and I have spent the rest of the day reading a book about silence written by a Buddhist monk. It's okay, a beginner's book, and it helped me to get back into a groove that I thought I'd lost, and I'm peaceful now. Peggy calls me to say her train gets in at eight, so I put some boots on and a jacket and dump myself into the van and go and get her from the station.

The cherry trees outside the front door are turning red, and soon all those thousands of little buds will open and people in the street will come out to photograph them. I planted one of them twenty years ago when my grandmother died; the other is maybe a hundred years old and is cracked and hollow and sheds limbs every winter.

A Vase of Cherries

Rain, then sun in cold air, and the cherry buds are pink and tight. I go out in slippers and cut a few twisted, knotty twigs with hanging buds and put them in a vase on my desk, so that I can watch them open over the next few days. I bend to my boots, the rain comes back from a darkening sky, I stay half-bent, midway through tying my laces, and wonder what to do. I wander to the back door, one boot unlaced. The cat follows, sits by my feet and we both wait and look through the glass. After about twenty minutes or an hour has passed, the sun comes back and I go to my van, and the cat comes through the front door with me, runs off into the wet grass and I drive to the garden that's waiting for me.

The lawns won't need cutting for a few days, and all seems quiet and ready for the sun to come and warm things up. There are buds everywhere – the trees and the tulips. The rain has cheered up the sparrows, who are singing and shouting like children playing playground games; wood pigeons call their sad, deep call; and the seedlings are well and a bright new seedling-green in the glasshouse. I water them and look around for a job that needs doing. Miss Cashmere's tortoiseshell cat joins me in the warm greenhouse. I bend to stroke its neck and it arches and purrs, jumps onto the bench and tries to bed itself in the seedlings, so I lift it off and put it in an empty seed tray; and it turns round and round, then curls to doze and occasionally opens its eyes to look at me through massive, dark pupils, then slowly closes them again. Today I am happy to have the company. Some of my closest relationships have been with cats; all of them have been teachers, and time spent with them is never wasted. Watching them luxuriate and do nothing but stretch out, I am reminded to let go and only use energy when I need to, to stay calm and still when nothing else is required.

The soil in the beds outside is too wet to hoe and the pruning is all done, so I go to the woods to see if there is anything that needs doing down there. On the way I enjoy the big twinkling sky and the little twinkling white flowers on the branches, and the few hardy bees that have popped out of sheltered places to sniff around the blossom, and I feel like I've trudged for miles and begin to feel a little guilty that I haven't done any real work today, but will still get paid. It's like I'm waiting for inspiration to strike me, like a poet.

Peggy says I work too hard and that I should slow down a bit, and the cat reminded me, so the inspiration comes to sit on a stump at the far edge of the woods and look out over the fields and hills, and the places like the ones I used to wander in, just like my poor Irish grandfather who went from place to place. He drifted like a wind-blown seed until he found a wife, my nana, who came from the Isle of Man, and who made him settle down and work in a fertile place where he could grow and send his own young drifting to another fertile place – and so on down the line, until I landed here in a fertile place of wet warmth that suits me for now, and from where I made my kids, who drifted off to other parts that suited them to grow. And so our vagrant connection is not with a particular place, but with a climate where there is warmth and moisture, and generous people who allow us to stay. I used to own nothing and have nothing to carry, so that I could drift, but now I have a house and a cat and other stuff, so I feel a little bit stuck.

It has taken some time to get used to being on my own in the garden this year after spending the winter days at home with Peggy, but the good old feeling of freedom is coming back, and I'm okay with it and being good to myself as I watch the crows come down from the clouds and land in the field, and wander about and then go back up again. I listen to the long grasses rubbing together in the tiny wind, and smell the green and the new leaves and the brown smell of soil, and feel the cool, damp air brushing against my bald head and the tips of my ears. Suddenly I feel cold because I paid attention to it, so I send my attention to feel the warmth of the sun instead, on my shoulder, where it touches. The things that we pay attention to get bigger and the things we ignore fade, and so the feeling of cold passes as I sit on this rotting stump. The small flying insects that disappeared over winter are starting to come back, and there must be grubs and beetles moving in the soil because the crows are stabbing at the earth. I think of the crows as 'for-ever birds', as if it were the same crows circling all the time, as if they were not living, mortal things, but something else entirely. I know that's nonsense, but I like the thought. On the other hand, I have never, in all my years, seen a dead one.

Dahlias

Last year she loved the dahlias (*Dahlia pinnata*), so this year I'm planting a whole bed of them. I ordered the tubers in the winter and they were delivered to my house in cardboard boxes and stored in shredded paper. I took them down to her shed, where they have sat in the cool, dry shade until now. Dahlias like very fertile soil, so I make a phone call on my mobile and spend the day driving the little tractor and trailer up and down the lane, a half-mile to the neighbours' stables, shovelling old manure that has sat all winter breaking down with its own internal warming processes. Urea in urine breaks down into nitrogen and, along with the fungus and the microbes, feeds the plants and breaks up the soil and keeps the whole thing spinning. That is why gardeners pee on their compost heaps.

I don't use farmyard or dairy manure any more, because herbicides sprayed to keep weeds down are part of the food chain now, and for a few years vegetable gardeners and allotmenteers have been getting damaged and failing crops, due to the carcinogenic glyphosate weedkiller that passes through the cow and remains in the manure. Glyphosate is banned in most of Europe and the Middle East because of this, but we still allow it here in the UK.

The stable manure is rotted, black and sweet, the clods made crumbly by freezing and thawing. She watches me from the window as I drive her little tractor and trailer.

By nightfall there is a pile of sweet-smelling rich manure as tall as me next to the compost heaps.

Today my shoulders and my lower back hurt from digging, but spring is really here and there is no time to waste, so I set to shovelling the manure into the wheelbarrow and push it across the lawn to dump it on the earth. The daylight is lasting longer now; it's still light, although there is a clear crescent moon like a sliver of God's fingernail poking through the sky, and I am tired and achy, and need to eat and cuddle up on the sofa with Peggy in front of the wood stove or the telly. In the small, inconspicuous tassel-like flowers of the beech tree as I pass, a bee is looking for nectar that she (for it is always a she) will take back to the hive to feed the queen, her mother. From the ground I can hear the tree is noisy with the hum of more than a thousand bees and, in the grass below, its strange kidney-shaped seedlings are starting to sprout, while above my head the bees already fertilise the flowers that will make next year's seeds.

Girlish

There are hundreds of kinds of dahlias, all brightly coloured and blousy and fun, that remind me of a glamorous 'salt-of-the-earth' barmaid who would have excited me as a boy in a 1960s movie, and who eventually comes to grief at the hands of a misogynistic gangster, perhaps Pinkie from *Brighton Rock*; or of Barbara Cartland in a child's party dress or layers of negligee, lying on a sofa dictating her romances, fully made-up with massive blue eyeshadow and fluffy white permed hair. My visual references date me. Dahlias are fun. Gaudy and lots of fun.

I take the roots to the prepared bed; they are dry and hard, like bunches of fresh new potatoes. They will become a mass of colour, all mixed in with the few lupins that remain and which are already showing new green leaves. I'm smiling – this is anarchic gardening, chaotic and informal, just the way I like it. While I am digging the planting holes Miss Cashmere comes wandering down the garden on her way to the summerhouse with her newspaper, and her previously clouded face seems to have brightened like the day, perhaps because of the blossom appearing. She says a very strange thing to me; she shouts, 'Swallows return, Marc!' And I look around and can't see any swallows. And she watches me and smiles, then says, 'Well, perhaps not yet, not here anyway.' She says, 'In Japan they have seventy-two different seasons, and this one is called "Swallows return".'

I tell her, 'I didn't know that, and I think it sounds rather lovely' and that 'I would like to know more'.

She says, 'Wait here' and, carrying her newspaper, she almost runs back to her house in an excited, eighty-year-old-lady way, all girlish with her knitted green dress stretching round her legs and her hair bobbing around. Well, she totters. She leans forward a little, then quickly she staggers.

And I smile and wonder what the hell has come over her. I wait until I think she has forgotten all about it, and I go back to planting dahlias, but shortly she comes back, quite a bit more slowly, carrying a thin yellow booklet with faded cardboard covers, which looks old and thumbed and so loved that it seems covered in life and bacteria.

'Here,' she says breathlessly, happily shaking the booklet at me.

And I stay on my knees and open the booklet's pages of very thin and rough, almost transparent paper with bits of straw in it, and kneel there by the new dahlia bed and read that ancient Japan had twenty-four seasons, with names like *pure and clear* and *grain rains* and *insects awaken*, and these were split into seventy-two micro-seasons, each about five days long. And I ask her what the date is, and today is indeed in a period called *swallows return*.

I can't see any swallows, and ask her if the descriptions work here, and she says, 'The climate is pretty much the same, I suppose', and then, 'I have no idea – keep it if you want it.'

And I say, 'Thank you very much', because I have just fallen in love with the idea of the little book and decide to keep an eye out for swallows.

In all the years, I cannot remember her ever giving me anything before, and I kneel there like a fool, smiling; and she is smiling back, and then the spell breaks and she looks around for her newspaper to take to the summerhouse and realises she must have left it in the house somewhere while she was looking for the book. 'Well, there we are,' she says briskly, and much more slowly she goes back to the house and doesn't come out again. She didn't ask me what I was planting.

I put the little book in the pocket of my old tweed jacket that is hanging over the wheelbarrow and go back to planting dahlias.

Love Is . . .

It is spring and the world is full of newborn things as the creative world wakens and expresses itself. The thrushes are collecting twigs and grasses. I trim my beard and throw the cuttings out into my little garden for them. Urban seagulls on the roof have paired off; they're waddling along, heads down in submissive gestures. Crows are flying in pairs as they go to roost in the evenings. Seeds are sprouting, leaves are unfurling, lambs are being born in the fields, daylight is longer than the night. The hibernating insects surface, and Miss Cashmere is full of the joys of spring. It's all-out birth and rebirth, and she looks cute with her white hair piled up and wearing a black dress, tights and flat shoes. At eighty-odd years old, spring hormones rage.

The poet Rainer Maria Rilke talks about young love consuming everything. He says that when we fall in love as youngsters, we lose our individuality and become at one with the other person. Our needs, our desires, hopes, wishes and dreams all become bound up in the beloved. We hope that love will make us feel complete – how we felt as tiny children before we developed a separate iden-tity, when we were just part of a mother that fed and bathed and nurtured us. But the minds of other people are closed to us, and soon such all-consuming love becomes aware of its isolation; we realise the other person is not the same person as us. We are trapped inside an isolated body that seems to end with the skin at our fingertips. The pain of separation begins again, and we start to wonder if it is really love at all; and if it ever was, why do we feel so alone? We try to fuel it up once more by getting married or setting up house, having children, to give the whole chaotic mess a context and prove our love, in the hope it may come back to us. But the fire settles and the heat becomes warmth, and awareness of our solitary nature returns and we have to figure out who we are once more.

All this childish wanting and fantasising is a hopeful trance. Every moment of it we move slowly further from reality, unable to see through its fog that we are already whole, connected and complete, as part of everything that lives and swirls and moves across the earth that grew us.

I was stubborn or cowardly enough to stick with Peggy through all the heat and cold and she remained by me. She had more than enough reason to leave, or have me not return. I was not always a good man or well behaved. She could not fix me or make me whole, because I was already whole, and she waited for me to remember that. I enjoy who she is and who she becomes every day, watching her being and doing what she does, changing like an expression of the creative earth, without any need to please me. I am already complete, I do not need anything from her.

There used to be sadness when she was not about. Now I allow any feelings that I don't want to experience to drift on by. I don't entertain them, and I keep the ones I want, I choose my trance. Only thinking drives people apart; we only disagree when we speak. When we do not speak, but share and enjoy the world in silence – when we do not plan, but experience – we are joined. There is no me or you, no us and them. Each day I can be newborn and, like a newborn, feel connected to everything around. You are not alone, you have never been alone, you will never be alone.

The Window Cleaner

This garden is my temple. I come here and expect to feel and taste the world. I make it lovely for the pleasure of it being so, for the labour that is good for my body and my mind. Here I just am, and here Miss Cashmere is. I think perhaps it is her temple, too. There is no chatter, we are mostly silent. There is nothing here to make us feel dissatisfied with life.

The magnolias that flower white on bare stems now show the pink and brown of fading cup-shaped blooms, behind the bright of new small leaves that lift the tree into another dimension. The tender baby leaves of ash are curled and tiny hands, the red of autumn leaves as they slowly pump in chlorophyll, unfolding them and turning them green – the reverse of what will happen when the darkness and the cold return, when the tree stops making chlorophyll and starts to break it down into smaller molecules and other colours show themselves. It throws its useless leaves away when we begin to celebrate with winter red and green. How much chlorophyll there is in the world; how much we all depend on that simple chemical converting sunlight into carbohydrates, which feed all living creatures and create the oxygen that we all breathe.

The hawthorns are already covered with tiny white flowers on the sunny side of the sheltered valley, where the gorse flowers that smell of honey and taste like garden peas are shining yellow and bright.

A robin watches me as I sit on the ground, pulling weeds from between the tulips. The sky is grey and very lightly drizzling, but I don't mind. I'm comfortable here. I have a foam pad as a seat and a warm tweed cap and jacket that repel small rain like this. In the rambling thorny stems the sparrows sing and fluff their feathers in this tiny pin-head rain. Behind them in the distance a siren wails as man goes about his business, too. My aching hand pushes the little three-tined fork into the ground beside where I sit, and my worn joints click as I lever out the dandelions and wild garlic that have invaded the bed and throw them into the bucket. I'll take the wild garlic home for soup.

Before I see her, I hear her coming slowly, in her waxed cotton stockman's coat that seems as old and stiff as she is, that creaks as she moves and nearly touches the ground. It's buttoned up and below it are the feet of old green wellies, dusty but starting to shine in places as the rain washes the dirt off in thin streaks. She talks about the monkey-puzzle tree (*Araucaria araucana*) at the front of her house that shades a bedroom window. 'I told him I wouldn't have one in the garden,' she said. 'I hate the things, but he planted it anyway, and look at it! And now I can't get rid of it – I think he did it as a joke.'

I could have said that her long-gone husband wouldn't want her to be unhappy. Looking in through her bedroom, as the tree does, was not, I imagined, his style, but I didn't want to make a case to cut it down. I remain quiet. I would

135

quit this place and live in hunger before I cut it down. Monkey-puzzle trees live for a thousand years and grow very slowly. I'm just a fly to that tree, which is worth far more to this planet than any number of us world-breakers; more rare than us pale grubs, it's as high as the roof of the two-storey house and, if it reaches maturity, its giant cones will drop seeds as big as almonds that taste of pine nuts, from between its ancient spiny leaves.

A moment passes and I realise that she simply wants to complain because she is in a complaining mood. She talks about the weather being grey and dark all the time (even though for the past three days the afternoons have been warm and sunny, and I have been able to work without a jacket on). Then she tells me about her car, which needs a service, and then complains about the moss in the lawn and about the window cleaner, who was supposed to come yesterday, but didn't. And I start off by saying that it has been lovely up until now and the car will be fine, and if it isn't, they'll fix it; and the window cleaner will probably come tomorrow. The pressure of her meaningless and minor complaints is like a river in flood, so I take to the banks and let the waters go by and say, 'Ah well, I'm sure it will all get sorted out.'

Her mood has been up and down recently: last week girlish, and now whining. She used to be very focused on the outside world, read her newspaper every day, but she seems to have gone into herself and isn't happy there. I wonder what has happened to send her into that territory. Remembering her lost life-partner perhaps, as the daffodils fade, or some new illness or pain has made her feel out of control in some way, which for her – who I imagine as always being in control – must be a frightening thing. Other people's inner lives are closed, so all I have is her words and manner to respond to. Sitting on the ground, I look up directly into her eyes and I say, 'The window cleaner will come'. Some kind of warmth seems to pass between us, and I feel something in my chest which is the same thing that I feel every time I see her face or her shape in the distance.

Then she says, 'Yes, he will, won't he?' And she pauses and wanders off to the summerhouse, where I see her in her wicker chair while her cat, as wise as anything, sleeps curled up on her lap, pondering as the rain drips off the roof and into a little puddle by the door.

The window cleaner arrives and I call, 'Good morning' to him as he brings his long brush and hosepipe through the gate, and he grumbles something. Gorse and hawthorn flowers by the roadside as I drive home, and in the fields there are lambs, as spring and Easter creep in.

Tulpen

I'm sitting by the back door with my notebook. The cat is curled next to me and has her head on my ankle. Peggy comes in and calls me a lovely old hippy, and I smile and don't say anything because I am feeling big and deep, resting here before I go to work. She goes back upstairs to her desk. She writes fat books of stories that she makes up in her head. I just write down what I see. She calls me her strange, enchanted boy.

The cherry flowers in my own little garden, which have been in bud it seems for many weeks, are opening at last. If I had the will, I could count the handful of open blooms, but within days there will be tens of hundreds. Today I will cut the grass and trim the edges and hoe the beds, and I'll get it all done as quickly as I can because it is going to be hot, and labouring out in the shadeless heat is no fun at all. So I plough into it like an ox and think of nothing, following the mower until, five hours later, my stomach complains and I stop and go back to my van for a lump of cheese and an apple, eaten off my pocket knife, and drink a litre of water to replace that which I've lost.

Miss Cashmere walks by. Eating in my van, I feel guilty that she has caught me not working yet again, and I hear myself explain that 'I've nearly finished cutting the grass and will edge the beds next.' 'Of course,' she says, as it needs no explanation, and she smiles at me as if I were a child trying to please her. And I feel even worse. A flash that could become resentment, if I let it, as I think: I've earned my lunch. Bloody working-class values creeping out from the depths, which get crushed as I return to finish my dance with the mower and bees and the sparrows singing.

Hoeing a bed of new tulips, so red – was it really accidentally, carelessly? – I cut off a fresh flower head and it lay there, flapping breezily. Looking at it held me tight and aghast. I look round to make sure that she hasn't noticed, and bend to pick it up and stuff it in my pocket, hide my shame away.

The massed tulips, pink and red, in this garden nod their turbaned heads. 'Too excitable,' Sylvia Plath said in her fabulous poem 'Tulips'. Written as she lay in a hospital bed, like a 'pebble' wrapped in bandages after an appendectomy. Bright and red against the white of walls and sheets, they disturb her peace with a noisy explosion of colour, while she wants to lose herself in the calm and feeling of utter emptiness. She writes of being watched by them while the nurses sail by like seagulls. I want to lie next to her, looking at the noisy tulips, another pebble wrapped in my own bandages, and sleep.

Tulips arrived from Turkey in the sixteenth century and went to Holland. Named after the Persian *dulbănd*, meaning turban, which they resemble. They unwrap before the summer comes, then drop their petals on the ground, leaving their naked green stems, moist stamens and leaves for all to see, which slowly dry out to brown over the weeks. Smaller bulblets grow on the sides of the mother bulbs and can be split off to make new plants. Many gardeners dig them up and put them in storage for the winter, but I don't do that here – there are too many of them. In my wanderings years ago I met a Dutch girl called Tulpen, who spoke very little English; I spoke no Dutch. Not speaking much was good. We just held hands and smiled and dropped our petals to the ground.

Swifts Arrive

I want to sit under the cherries in the sunshine, drink whisky and hold them close until May, when the wind will come and blow the tons of weightless pink away. I crave to be lazy, to plant my feet on the earth for days or weeks, to wander.

I have no land, no square mile. I carry no history of place, just vague memories of a few towns and villages, a forgotten endless passing train of small, cold houses. I was born in a place I never went back to. I'm vagrant and my culture is drawn from indigence; I have a passport for a country I don't feel part of. I have no master and there is nobody I'm willing to be master to – an anarchist. My bloodline stories are the songs from hungry mouths that have, like mine, eaten cardboard and leather and sheltered under the lees of rocks and inside poor, unheated houses, and shovelled coal into boilers in factories and locomotives and mill engines, broken mountains into stones for railway lines to rest on, moving as the job moves on, following the tracks as the tracks go to wherever they go.

The men like myself have fought and been beaten and sung drunken poems and songs in bars with sticky wooden floors, cried then turned round, smiling to other men who know, generously pretending that they haven't seen. Women have spun cotton, been battered and abused, cried over the hunger of their babies, the unpaid rent, the drunkenness and inconstancy or illness of their men, and walked the empty streets at 4 a.m., arm-in-arm with other girls and singing. My transience fills me with the deepest feeling; a truth that roots me through the mud all the way to the Earth's hot core, I'm mud. Like mud, my wandering has slowed and stopped, I have a dry, warm place to live and would rather like to hang onto that as I age. I would like to find a comfortable chair.

I am working the land. The many jobs that I did before this were only a replacement for working on the land. Bosses tried to make me do what they wanted and believe their beliefs, but I was a wild patch of thistles and nettles and butterflies, and I did what I wanted instead. I left the factories and workshops and the double-edged rules and slick bullies, and came back to the place where I belong, the land that sings me as its song.

A swift flits across the meadow, then another one and half a dozen more. I am filled with joy and stand to watch them circling. They come for a few weeks to breed. They arrive in waves, some heading further north, some staying in the village to lay their eggs and feed their young under eaves and in the old bell tower, and then they leave and never stop.

Song

I fell asleep to the blackbird singing outside my open window and woke to it again. The first sign of dawn. 05.16, there's no light in the sky. My cat is sleeping on my feet. Now the urban seagulls cackle on the rooftops, waking late and noisy, backing up the sound of hedgerow birds. A wood pigeon comfortably tells me he is there. It's not yet fully light.

Peggy is away for a while and I am like a balloon let go. I lie about until the day is full, and in the cold of morning I get dressed. She is a writer and, like other writers, she travels: to see her publisher, to read at events across the country, book festivals and the like. She has become a Gipsy like me. I can still hear the brass tongue's soft click in the lock as, otherwise silent, she left in the early hours while I drifted awake. It has rained for two days and the river is about to burst, I'm holding it back. Now soft rain makes the earth smell of rotting leaves, growing fungus, something sweet. It patters in the background behind the excited songs and calls of many birds. I stand by the open back door, cat at my feet and cold air on my arms, and watch the cherry branches weighted down by a couple of thousand pounds of feather-light blossom. The knobbly bare stems, whose shadows only make their shining wet the brighter, touch the earth and I long for her to come home. I let the wanting pass, leave the house and drift into the warm and scattered, spotty rainy day. I feel it on my skin as if it has never happened before.

I stop at the supermarket and buy a lump of Cheddar for my lunch. While I'm there I pick up a packet of lettuce seeds from the twirly wire stand by the magazines. In the queue in front of me a boy is buying football stickers with his pound. Outside a fat little girl in sunglasses is doing the Macarena in the bus stop. The rain stops and the sun breaks through, the road steams a little.

Miss Cashmere's car is there, but I haven't seen her for a few days and I find a note on the greenhouse bench. I've no idea how long it's been there. She leaves one from time to time. It is in an unsealed envelope, in blue ballpoint on small and matching, age-yellowed, old-lady writing paper, and in handwriting that is full of tall, right-leaning characters with shaky loops it says:

> Dear Marc. I am going away for a while. If you could keep an eye on things while I am away that would be very helpful.
>
> My regards
> Dorothy Cashmere

In front of the greenhouse there are three raised red-brick beds. One is the strawberry patch, the other has raspberry canes and in the third I sometimes plant a few simple vegetables, things that are undemanding: lettuce, radishes, kale and chard. A few beans. Today I draw two drills in the tilled surface with my finger and plant half a row of lettuce (*Lactuca sativa*) seeds.

We have got into the habit of sharing these few crops; she likes the strawberries and I often take a handful to have with my lunch. I tend to take home most of the leaves of kale and the chard, when it comes through later in the year; she doesn't bother with it so much, and my own garden is too small and shadowed to grow much. Her ginger cat mewls around my feet. I wonder if somebody will come to feed it. There are plenty of birds and mice around here, but I think it is more of a house-cat and not much of a hunter. I feed it some cheese from my hand and it seems to be starving. It doesn't know how to live in nature, I am a wilder creature than this cat. 'It's just me and you now, cat,' I say and it turns round and round in the strawberries as I sit on the edge of the raised bed and stroke its back.

I bring it food each day for a while, and it runs to me when I arrive: a tin of Peggy's tuna, my own cat's tinned and soft food. After three days of feeding her cat, I see a small van drive up and park outside her house. A young woman in an overall gets out, uses a key and goes into the house, then comes out again. 'Have you seen the ginger cat?' she asks me. 'Only it hasn't been eating its food.'

'It's down by the greenhouse,' I tell her. She throws the uneaten cat food in the bin, washes the bowl and puts out fresh food and a bowl of water. The cat has already eaten a tin of Peggy's tuna today.

I have decided to call the cat Malcolm. Malcolm keeps me company and is the only person, apart from the cat-sitting woman, that I speak to all week. 'Come on, Malcolm,' I say, 'help me choose a few strawberries to have with my lunch.' Malcolm wanders off towards the meadow. Malcolm is my cat now.

World Sings

The cherry blossom that hangs in great clumps from the ends of knobbly, thin branches of the ancient tree is so heavy that the outer branches bend to touch the ground, like an old lady resting her feet on a footstool to take the weight off her legs. In this ripeness is its end and some blossom has started to fall already; the rough black tarmac road beneath the trees is scattered with white and pink polka dots turning brown.

Like the blossom, we do not come into the world from the outside; we are born from the world, the world expresses us in the way that music is the expression of an orchestra or a song escapes the blackbird, like a wave that flows into and out of being. In my world there are no natural things that are not connected at the source. If there is a God of any kind, it is not above us in the sky and looking down, but down below us, blowing us out and playing the world like a flute; the pine tree, the ginger cat, the boy with his football cards, the Macarena girl in sunglasses – all emanations.

After a week of sunshine and taking it easy in the garden while she has been away, it's time to get back to doing some work. From now on, mowing the lawns will be a weekly task until the autumn comes. Later I wander along with a pocket full of calendula (*Calendula officinalis*) seeds that I gathered last year and left to dry in paper bags in my van over the winter. I collect them every autumn, pockets full of them, a variety I planted many years ago called 'Art Shades'. These happy and pretty little annual marigolds, in peach and straw and orange, have just one function in life; like all the other plants that live for only one year, they make as many seeds as possible before the year ends.

There's a little sunny bed by the fence on the left where the golden calendula grow mixed in with blue cornflowers (*Centaurea cyanus*). When the calendula flowers, the bees come and pollinate them as they fly from flower to flower, attracted by the scents and sun-like petals, which form a landing platform spreading from the centre of the flower; sometimes they have ultraviolet markings that insects can see, but humans cannot. Flowers are a plant's reproductive organs, and I remember my school biology lessons where I learned the fascinating names and functions of all its parts: the petals that attract the insects with colour and perfumed oils; the stamens made up of anthers that are covered in pollen and sit delicately on the end of the filament, which together stamp pollen onto the bee's hairy body as she wanders around the flower.

The bee stuffs pollen into big baskets built into her back legs that can weigh half as much as the bee itself when they are full. The pollen and the nectar are taken back to the hive for food and are shared by the whole colony.

As the bee wanders around, pollen caught on her hairy body sticks to the moist stigma in the centre of the flower and travels down the long tube of the style, which leads to the ovary, where it fertilises the tiny white embryos. There they will fatten and dry throughout the summer and into the autumn, when the flower loses its petals as it no longer needs to attract pollinators and the bees go into hibernation, having done their job for the year. I'll collect some of the seed and use it to fill bare patches, and throw some into unloved gardens and patches of bare soil and grass verges wherever I go. I am signing land that isn't mine – profligate, licentious, promiscuous – and my babies are everywhere.

Love-in-a-mist (*Nigella damascena*) seeds, small and black and roughly cornered, collected from last year's plants and cast among the columbine (*Aquilegia vulgaris*), are shooting up towards the taller leaves. The columbine are also known as 'granny's bonnets' and 'doves'. This pretty and common cottage plant is wild, breeds freely with other colours and varieties of the same plant and its children rarely look like the parents. I love their promiscuity and let them grow wherever they appear. Their flowers seem a circle of tiny doves perched, facing inwards, touching beaks; 'Columbine' comes from the Latin *columba*, a dove, 'Aquilegia' from the Latin word for eagle, as the flower appears to some people like an eagle's claw. Hawks or doves? It's just a plant, and every thought we attach to things is ours.

A Broken Heart

In the shrubbery the little bleeding heart (*Lamprocapnos spectabilis*) that flowers every year is dripping with its sentimental pink, heart-shaped flowers, each with a little drop of 'blood' hanging below, which makes me shudder with its horrible pinky-hearty tackiness. As I pass the bleeding heart I wonder about my emotional response to this silly little heart-shaped flower that I dislike out of all proportion. I could quieten my mind, let go of the judgement and move on, but, wondering why I judge, I let the thoughts come and realise that what I respond to is the belief that there are 'the kind of people' who would like this little hearty, trashy thing, and I judge their sentimentality.

I have always had a problem with sentimentality. My father was a sentimental man, as were those he hung around with. Partisan men who proudly thought themselves 'MEN!', who loved their mothers but believed that other women required domination, that boys had to be brought up hard, and girls were adorable and in need of protection; men who worshipped their own 'team', but could not applaud the team they lost against; who would argue that they were Christian and believed in a God, but were too proud to pray and would happily abuse and mock the weak, and would send their crushed wives a card with a pink heart on it on Valentine's Day. These men fawned over their heroes and aspired to be like them. Men who believed in the authority and power of those further up their chain, and who would do their bidding when they rattled the links; who understood and accepted their place and believed in their own authority over those men and boys further down the chain. 'When I say JUMP, boy, what do you do?' The poorest and the unemployed lifted weights, so they looked powerful, and had aggressive-looking dogs to control. Women and girls were not even on their chain, unless they were 'one of the boys'; they occupied a pretty little silver palace of their own, like a charm on a bracelet, dependent and sweet and grateful. Anybody 'off the chain' confused the hell out of them and was considered to be the same as the women, except without the silver palace. I was called a 'poofter' or 'gay' at various times for being a vegetarian, for not playing foot-ball and for reading books.

I realise, and understand, the conflict in these men, despite their laughter and brief joy when they were in drink, or when their team won or their child scored a goal, and despite their anger and occasional violence when they lost or failed. They have a pride that is insecure and contin-ually shaken as they jockey for position in the hierarchy, even with their closest friends. I feel a deep sadness for them. For their conflict and insecurity, for their broken hearts, for their distance from the things they profess to love. Their lack of peace and happiness defines them just as much as their passion and strength; they do not have the power to not express the power they have. Such men will rejoice at causing the end of the world.

This little, highly toxic plant, the *Lamprocapnos*, is a native of Siberia and China, where so many of our British garden plants originate from. Its name means 'two-spurred' because the little heart-shaped flower is actually made of two separate petals that each have a tiny spur on the end. This passionate heart is not unified; it is broken and each separate half has little hooks. The plant tells me that sentimentality is about separateness.

I understood that I had unwittingly considered myself separate, too, by judging it as 'tacky' and making assumptions about people who might like it. 'For there is nothing either good or bad, but thinking makes it so,' said Shakespeare in *Hamlet*. Two thousand years earlier Epictetus the Stoic said, 'Men are disturbed not by things, but by the view which they take of them.' I decide to love the little plant and not be disturbed by its sentimental appearance but to try instead to understand the inevitable sadness of passionate, disconnected men and to hold close the knowledge that we all suffer, and love, in different ways.

Mouse

My cat brings in a mouse that runs across the kitchen floor to hide under the stove. Cat wanders off and curls on the sofa and opens an eye from time to time to watch, to see if her mouse has come out again.

'Oh, you bastard!' Peggy says and then, 'Go and get it, Marc – get it out! Bloody thing!' So I crawl around on the floor, looking under the cooker, and I can see the mouse in the corner, looking at me with its big round black eye; it's hunched up against the back wall. I try to reach in, but the mouse is too far away and my arms are too big and don't bend in the right places. I don't want to hurt it, so I go for a broom to scare it into moving and the bloody cat sits on the sofa, all relaxed and watching me, while Peggy stands behind me saying, 'Have you got it, have you got it yet?'

I'm starting to think this is between the cat and the mouse; it has nothing to do with me and I'm interfering in their relationship.

Mowing in the Rain

The birds sing and Byron, the jobbing gardener, is cutting the lawn in the communal garden nearby and stuffing his mower. I can hear his mumbled curses at having to come out to earn his money, he's kneeling and sticking his hand in to pull out the lumps of grass that jam up the machine. You can't cut grass well that is wet with rain, but the desperation of the jobbing men to earn their daily rate is enough to make them do it. I've had my time doing that, ripping the grass instead of cutting it, getting drenched with raindrops the size of marbles sizzling off the hot exhaust. But I'm comfortable now, with a comfortable job that pays me every month to keep the garden looking good. Even in the winter, when there isn't much to do, she pays me, so I won't go away and start working for somebody else. I can hear Byron going up and down, and the mower struggling as the wheels make muddy grooves through the lawn, and the heavy clumps of wet grass get jammed in the chute instead of flying into the collector on the blast of warm air that comes from the blades' back-draught. Poor bastard.

Nearby there are trees where in the evening I see bats and hear owls. I live in the remains of a woodland not far from a river, where I like to stroll when I'm too tired for the mountains. I am not designed to be indoors and feel dusty and muggy, but it is too wet to work today, so I put on a raincoat and head out to air myself. 'Morning, Byron,' I say.

'Morning, Marc, bloody awful day for mowing,' he says, 'but I've got a backlog of work that I couldn't do in the storms last week, and the grass keeps on growing.'

'Sure does,' I say and again wonder about this dreadful obsession with cut grass that home owners seem to have. 'Many more to do today?' I ask.

'All day, mate – all day long,' he says. 'I'm sick of it. You not working today?' he asks.

'No, I'm taking a day off,' I tell him. I don't tell him that I don't go to work in the rain. I wouldn't want to rub it in.

Under the trees, by the river banks and hollow stumps where tiny ponds in the depressions between their ancient roots are collecting leaves, the first croziers unfurl as ferns and bracken make their bids for damp woodland dominance, each stem starting as a hairy green spiral; and slowly, as it unrolls, the leaves curl out from the sides, alternately – one on the left uncurls, then one on the right, as the stem itself unrolls. Each little leaflet is a copy of the larger stem. Nature likes to play with rhythms and harmonies.

Further out, where the shade is not so deep, the plantains (*Plantago lanceolata*) send up flower heads from between the strappy leaves, a furry brown lamb's tail on a slender green stem with a circle of pale-yellow flowers as small as a pin-head, on thread-like stems opening around the bottom of the bulrush-like flower head. And then another few above. As the lower ones mature and slowly fade to brown, the ring of flowers works its way up until there is a crown of flowers at the top and the ones below are ripening into a hard rat-tail of green seeds, browning in the wind. I know they open from the bottom to the top, because I visited one every day and watched it for about two weeks.

The umbellifers hogweed (*Heracleum*) and cow parsley (*Anthriscus sylvestris*) are starting to open, western water hemlock (*Cicuta douglasii*) too – the plant that Socrates, among many others, took to end his life. Giant hogweed is here as well, and brushing up against it will cause no immediate harm, but when that bare arm is exposed to the sun, the phototoxic chemicals can form blisters on the skin that look like boiling water has been spilled from a kettle. The giant hogweeds are down by the water where nobody but dogs and foxes, ducks and herons go.

The nettles are lush; they taste like spinach and the young leaves make a decent soup. In Celtic countries like Wales, Cornwall and France, they are sometimes used to curdle cheese. In Dorset there is a world raw-nettle-eating championship, farmers feed them to chickens to make the egg yolk more yellow, they attract butterflies and are a sign that the soil is very fertile. Their sting is said to be a remedy for rheumatism, and over the years I have come to enjoy the unexpected burn of their microscopic crystalline hypodermic needles injecting me with histamine and serotonin. Even the sharp stings of wasps and bees, I'm told, have health benefits and don't really bother me that much any more.

As the daylight fades and the showers end, I head to the cathedral. Behind spiked iron railings two eye-pupils' width apart – eye spikes – a field of the rich dead rests, and massed blue stars of forget-me-nots strain to escape their green leads. The millions of white candles of horse-chestnut flowers light up in the falling sun, and the frothy field of cow parsley catches the light, as the water drops on their flowers send sharp rays of sunlight out at eye-level, enough to make me shield my eyes. The humpback bridge over a dry ditch to fields of stones is washed in river mist. Shouts of students playing soccer on last year's meadow. A dead stone-winged soldier looks down from the cathedral roof, head bowed, leaning on his sword. The red stalks of Himalayan balsam hang over the river and reflect in the slow, brown water. I can smell its perfume in the thick scent of rain and river.

Roses grow around my windows; they are wet and glowing against the darkening sky. Leaves and rain. The old lady who lives at number thirty-two totters by on the fallen blossom, unsteady because her free hand is being used to carry an umbrella instead of a stick. Dripping, green and sloth-like on the knobbly branch of an arm, her handbag bulges. A man from down the road who always ignores me when I say 'Good morning' hurries from left to right with a bottle of wine. People are starting to pass. They are coming home from work.

Floating Islands

The winds came last night, as they do every year when the cherry has blossomed. The wet road looks as if a child has dotted it with a clumsy brush and pink poster-paint so thickly I can barely see it shining, all tarry between the blobs. The gutters and street drains run with rivers of floating petals, a regatta that clogs and blocks the gratings so that the water rises and floods the road as it tries to reach larger waters behind great dams of soggy blossom. The wind continues to blow through the morning; downwind, the street fills with a sticky unmelting snow of petals cast by thrashing branches. Warm and sheltered, I am like an eager child watching snow fall and waiting, desperate for everybody else to wake, so that he can go out to play. This happens every May, and here on its edge the wind and falling blossom are more reliable, more predictable than real snow ever is.

In the evening Peggy and I decide to go to the Forester's Arms for a drink. She'll have a glass or two of wine and I will have a couple of whiskies and then we'll head off home again. We sit on an empty wooden bench by the window and watch the customers coming in and out, and see a few people that we know enough to say hello to, but not so well as we know what they are called; and others that we know enough to say 'Hello' to and 'How's life?' and who we do know what they are called, but are not close enough that we would want to sit with each other. And after our drinks, as we are about to leave, a man comes in carrying a pair of black cases and then goes out for some more, and then a stand for a loudspeaker, and I say to Peggy, 'It looks like there might be some music' and we decide to stay and move into the main room.

Byron is in there, wearing a dark suit and a white shirt and dark tie, and we talk about how there looks like being some music, and he says that he was just about to leave, but that he 'might stay if it is songs from the Sixties, but probably not'. The singer sets up his equipment and starts playing the backing track that he's going to sing along to, and he's singing songs from the Sixties and Byron starts pretending to sing into an imaginary microphone; and I ask him if he is a bit of a singer, too, and he says that he is.

I end up drinking lots of whisky and dancing with Peggy, and then with some nurses and then with the land-lady, who tells me that she is feeling better, but still feels like crying sometimes. I don't know what she is talking about, because we haven't been in there for a while and she seems a bit delicate, so I don't ask her, because I don't want to set her off crying or make her have to talk about whatever it is that has been making her sad. So we dance and I whirl her about, and she seems a lot better. I am not a good dancer, but I am apparently very enthusiastic. We stay until the pub closes, and after that Peggy and I are more than tipsy and we stagger home, trying to hold hands, but we are so drunk that we keep veering away from each other and losing our grip. We laugh about how drunk we are and about what a lovely night we have had, and how we should go along next Saturday to see if we can do it all over again. By the morning we have decided that perhaps we won't do it again next week, but perhaps the week after, maybe, if we still feel like it.

May

Peonies Bloom

My elegant Russian neighbour walks through fallen blossom. Chasing off the seagulls who are courting on the pavement, rubbing beaks together, gathering moss, pulling lumps of grass from lawns. They are building nests on the flat roofs of the houses. When they've laid their eggs, they'll dive-bomb the pedestrians and children and dogs who live in the street. Then she'll carry an umbrella to protect herself. She hates the gulls. She has an old plastic carrier bag in each hand, which she has reused so much the print has worn off. We do that here, where beaches and rivers boil with abandoned plastic. Through the thin plastic I can see that she has bought bread and tins of beans, a cabbage, bacon and vodka. She is strong and handsome, dresses in elegant clothes; sometimes there's a slight Gipsy flourish to a blouse or a coat. Her scarf of brown printed with bright-red peonies around her shoulders blows in the breeze.

Peonies in sunny spots are open and some of the roses, too. I wonder if the Russian lady knew, when she chose her scarf that morning, that the peonies would be open, or if by some strange coincidence it just seemed the right choice. I wait for the peonies and pamper them, but the next heavy rain will be the end of them; their big papery petals will be knocked off and lie browning, looking as if somebody has carelessly dropped a box of small sheets of Japanese writing paper and is too lazy to clear them up. The Japanese love these flowers, and many tattoo images of them on their bodies. The peony tattoo symbolises the transitory nature of existence. Peonies have great big fleshy roots, and if you dig them up to move them somewhere else, they can take years to flower again, but left where they are, the flowers will become more abundant. Despite their glamour, they are home lovers, strong underground and hang on to their patch of earth for years, anchored like a claw in flesh, slowly growing bigger and fatter, holding on tighter and flowering for only a few days every year to earn their place.

The lawn as I mow it is full of daisies (*Bellis perennis*), and the 'day's-eye', like the peony, closes at night and then it looks like a tiny pinky-white pearl floating up on a string. These simple pretty (*belle*) blooms mean 'innocence' in the language of flowers and, like the peonies, the petals have been used to make tea; the astringent sap was used to heal sword and spear wounds; and its name perhaps derives from *bellum*, the Latin name for battle – the daemon of war.

Ophelia drops the daisy to the ground, symbolising lost innocence. As a young student, I fell in love with Ophelia in Millais's gorgeous painting in the Tate – a post-card Blu-tacked on my bedroom wall. Ophelia, floating down the river in her watery satin dress, red hair trailing, clutching a bunch of flowers – the remains of those she had handed out in *Hamlet* – driven mad by loss, and loss of love and grief, her innocence gone, her belief in the

innocence of her lover destroyed, she is literally deflow-
ered and destroys herself.

My mower passes over the daisies and leaves them
where they grow, close to the earth; the whole lawn is scattered with their pretty white heads. The blades slice off the
dandelion heads, upstrained against their tethers, leaving
only those that grow on shorter stems, safely tighter to the
earth, so that they'll seed; and only those low-growing
flowers will reproduce and perhaps change the appearance
of the species.

In the borders lily-of-the-valley (*Convallaria majalis*)
are showing through in clumps. Last year on May Day I
had gone wandering to northern France, visiting family.
Lovers hand each other *muguet*, lily-of-the valley, bought
from children in the marketplace who, watched by parents
from distant shop doorways, wander around selling
bunches for a euro each. These highly poisonous flowers
slow the heart rate and cause arrhythmia, but have been
found to act like progesterone and stimulate lazy sperm to
swim. It is all sex and death, where living things are
concerned. The big ox-eye daisies (*Leucanthemum vulgare*)
and the marguerites (*Chrysanthemum frutescens*) are just
starting to open in the meadow, they remind me of Peggy;
they are pretty and strong and Peggy is a nickname for
anyone called Margaret.

Gulls Rip Grass

It is 05.30 and this dawn chorus is an echoing thing of many small birds, which arrives in a wave. Then another stronger wave of new sounds, as people rise and flood. A throbbing truck rolls, then dims. Limpet-like, I cling to my home and try to resist being washed away into the separateness of the day and the drizzling pink sky. I'm reading beside the open back door, lyrical words in French that make me work hard for their meaning because my self-taught French is poor; Rimbaud's *Une saison en enfer*. The rain is here and I am glad of it; my back and the dry earth are glad of it, too. I've had enough of work, I'll stay at home.

The cherry trees are bending under the weight of water, their blossom is falling, browning and fast sliming the path, so the old ladies totter. A splash of heavier rain passes by as large drops bounce off the shed roof. Another in a higher key, as it hits the tarmac. The clouds dissipate to nothing. A second-rate sun takes the stage, the fading wisteria drips. How I would hate to leave the earth, the smells, the damp, the warmth and cold of changing air pressures. When I have been separated from the land, from the cycles of nature and the weather, working in factories or offices, I have become distanced from my own mortality, from feelings of being a human being in the world. I've become at times like a machine, programmed

to produce and to consume, and my behaviour and atti-
tudes became polluted and strange to me.

Still against the grim sky are the cherry trees. The
bad-tempered man from down the street passes by. He has
no shadow. He is wearing jeans. I'm watching blurry bright
colours moving through the wet glass and pondering the
possibility of going out. Maybe I will just stay here,
watching the big, heavy drops splat off the slate onto the
tarmac. Oily slicks of blue stuff. Mimi, my cat, curls next
to me and we share some inner silence.

A pair of bullfinches by the roses. I am old and have
to sit down to put on my shoes. That is how it is. Today I
am staying out of the fields, I have left the woods to the
crows and I am staying at home to write about it. In my
memory it is different. I think I am fit and healthy and still
young. I forget how tired I am and, as I sit and rest, I feel
exhausted, my legs ache and my chest feels tight. There is
a melancholy, a sadness to getting older; life changes and
there is a clear knowledge that things that have never been
will never be. But there is also a freedom that comes,
responsibilities are lessened and there is space to be more
creative. Less is possible, but so is more; age is a knife that
cuts both ways. I am perhaps at the most creative part of
my life now, and I am better at the things I want to enjoy.
Life has become hilarious – we laugh until we are nearly
sick sometimes and don't care about stuff any more. There
are fewer material things in my life, and I enjoy each of
them more than before. We are happy and confident and

able to cope with adversity. I meditate every day, an hour or two sometimes, and that makes me happy.

Nobody has passed the window for hours. Gulls rip grass and take it to the roof. Time passes and darkness falls. A jackdaw watches me from the fence with its hard, pale-blue eye that is magnetic to my own. The gulls are undisturbed. The winds blow as they do in May, and the remaining blossom rips from the branches and blows down the street and big rain comes. Heavy floods fill the gutters and I watch the storm slam into the street and listen to the wind as the darkness wanders over to join in. Storms are good times for me. The storm persists for three days and I stay indoors, drink tea, read and enjoy its passing and know that I am blossom.

Holy Thorn

A fearsome storm has passed. A cloud of steam from the central-heating boiler floats past the window, and through it I can see a seagull flying low. I am waiting. I have things to do today. I don't feel like doing things. Doing things involves dealing with stuff, and I'm fed up of the burden of stuff and doing things, and have been disposing of things I don't use and making my life simpler, more mobile. The ground is still wet and I decide to take my time. Later, as I dress, I can hear the children screaming in the playground; and as the clouds drift by, leaving the sky clear for the sun to come and warm the earth, I go out to do my work – just a few hours, I think. I leave the house at about eleven, after three rain-days off. There is still a little cherry blossom on the trees, but the leaves are more prominent than the flowers now, there is far more green than pink. The few pearls of petals will not last much longer as the pink canopy becomes green.

There are no cars parked by her house. I bend to the weeding and the unexpected sun warms my back. So quiet. So very quiet. One of those strange days when even the birds are quiet. Making no more sound than falling snow or petals. The powerful perfume of lilac hangs across the hot, still grass and the wisteria droops in great pendulous bunches on the wall, like giant bunches of small lilac grapes. The columbine is tall on its red stems and is tentatively opening a few small first flowers. The big African daisies (*Osteospermum*), too, are open and spreading their lank stalks on the ground, with their purple faces open and tracking the sun. All the colours are related to lilac and purple, and even in the shade the great brutal swathes of Spanish bluebells (*Hyacinthoides hispanica*) share their colour. It is all matchy-matchy, and round the corner I'm happy to see, among the new red leaves, yellow roses on the fence starting to open, too, although yellow and purple are not harmonious together and need to be kept some distance apart.

The blossom of the may tree (*Crataegus*) is open and its sweet, slightly rotting meat-scent will attract millions of insects, which will pollinate it and turn the flowers into deep-red haws. The birds like to nest in the hawthorn, protected by its dense covering of sharp spines. Nearby there is blackthorn, too, which is almost identical, but has already flowered and lost them. We have used these native plants for hedging for thousands of years and there is much mythology attached to them. They are guarded by faeries and worshipped by Druids. The Holy Thorn of Glastonbury supposedly grew from the staff of Joseph of Arimathea, which he thrust into the ground. The ancient and much-vandalised little tree that stands on a hill over-looking Glastonbury Tor is supposedly a descendant of the original. This particular variety blooms twice a year, once in May and again at Christmas. Cuttings have been taken and planted all across England, and a sprig taken from a descendant of this tree, growing outside St John's Church in Glastonbury, is traditionally used to decorate the Queen's breakfast table at Christmas, although the plant is considered unlucky and few Celts would bring it into the house.

It's a tree I'm particularly fond of. It has been used in herbal medicine to settle an irregular heart and to strengthen the elderly, and the leaves are known to many children as 'bread and cheese' and have been nibbled on for generations. It tends to bloom around the second week in May. In the past it would have bloomed at the very beginning of the month, but in 1752 the Gregorian calendar was introduced in order to make Easter fall around the equinox. May now occurs about two weeks earlier than it did then, but in recent years, as the planet warms up, the may tree has been blooming closer and closer to the beginning of May, re-establishing the older order of things.

The may trees are flowering early again, and I am watching a flock of long-tailed tits harassing a lone tree, dashing in and out, tiny little pretty birds that look like little fluffy grey balls with a long, stiff tail sticking out. Sometimes I find their abandoned nests after the chicks have flown; little sleeping bags, soft and warm, made of cobwebs and lichen and lined with feathers, with an entrance hole just big enough to put my finger in and feel how cosy it is inside. Every time I find one I crave a man-sized version, attached to the branches of a tree,

which I can pull myself into and sleep and wake and watch
the world go by, with my face looking through the entrance
hole like a giant chick. I smile and imagine a human-sized
bird coming to put food in my mouth.

Mercedes

I'm cutting the grass along the verge in the lane at the front of the house, using a petrol brush-cutter and wearing a hat and glasses and ear defenders to protect myself from the noise and flying stones. Every now and then a car comes by, but of course I can't hear them coming, so I have to keep stopping to look up and down the track that curves away in both directions. I can't see very far. I am covered in grass cuttings and step back to look, and a man pulls up in a black Mercedes. He is well dressed and there's an equally well-dressed lady in the passenger seat. I'm wearing a broad-brimmed hat, a beard and dark glasses. I didn't realise that I was in disguise until I saw myself reflected in the shiny car. I think I recognise the driver, but I am not sure where from; it could have been TV, but I don't watch much TV so I can't be sure, or maybe I saw him in the pub a few times. He leans across his passenger's lap, I pull off my ear defenders and he leans towards me and shouts a girl's name, and instead of grass and petrol fumes I smell warm, mixed chemical vapours of car air freshener and perfume and spray-on deodorant. He asks if I know her, but I don't, so I suggest that he asks at the farm down the road. As he drives away, I think about his nice suit and shiny car and the well-dressed lady with smooth skin that is the same colour all over her face, even on her cheeks and ears. But what I think keeps on changing.

I am for a moment uneasy, in my dirt and scruff and worker's clothes, and I am big and ungainly and feel self-conscious, and start to wonder if perhaps I am envious of his apparently wealthy and possibly glamorous lifestyle. I think that I am not, because I am happy in what I do and how I do it. I have nobody telling me what I have to do, and there is nobody that I have to tell what to do, and that is just the way I like it. Nevertheless I feel like a bumpkin, unsophisticated, and start to wonder about the value I have put on such things; how being sophisticated somehow seems to be better than being a yokel, as if I am somehow less than he is. I carry on with the brush-cutter, throwing noise and small stones and slashed weeds into a wet, green scented cloud that surrounds me. Inside there is peace while I ask myself what I mean by 'sophisticated', and realise that I mean glamour and clothes and parties, power and money perhaps. And I start to think that those things are all about fitting in with a society of some sort, buying the right things and competing to stay on top, and so inevitably I begin to feel that being a bumpkin is a much more desirable thing to be. I imagine him having a meeting in his office or having powder put on his face in the studio, being pampered, or buying a new car or being fitted for a suit, and I think: well, that would not be so bad, either. It's all swings and roundabouts.

Life is so impossibly fleeting, and money is immortal. The collection of possessions helps us to pretend that our lives go on. We can, if we work hard and fortune smiles on us, build something that can outlast us. But such things as we leave are usually broken up and reused to help somebody else feel immortal. The constantly cycling world smiles down on thoughts of permanence. In 'On the Shortness of Life', an essay by the Roman philosopher Seneca, the Stoic argues that life is not short, but that we make it so by squandering it on heedless luxury, being 'gripped by insatiable greed' or 'laborious dedication to useless tasks'; judging, or being judged, and dreaming of leisure and fun. He argues that much of life is not life, but merely time. Living as if we would live for ever, guarding our money viciously, yet squandering our time on useless activities and obsessions: arguments, wars and battles, politics, sorrow, and working to build wealth, while making plans to retire one day so that we might please ourselves and start living life properly, as if a future were guaranteed and death was not. 'They know their time is limited,' he says, 'but they continue to complain and carry on and would not change even if they lived a thousand years. They let their lives slip away as if they were worthless.'

An hour later the Mercedes glides by the other way, and the driver waves and smiles and his lady waves, too, like he found what he was looking for at the farm. And I smile and wave back and we're good with each other, and I like them and feel that they like me.

It is sunny and warm, but when the sun goes behind a cloud it suddenly feels very cold and I remember my imaginary man-nest in the hawthorn tree.

As I drive home, a man in a van drives in front of me, beeping and shouting because I am driving too slowly in my little van, and I feel threatened, as if my happy little bubble had burst and horrible reality had come in. He pulls up in front of me and puts his hazards on, gets out of his van and starts shouting in the road at me, but I do not know what he is saying as I keep my windows up and doors locked, and I feel for the wood-splitting axe that lives behind my seat. He gets back in his van, after ranting for a bit, and drives on, and I feel that there have been far too many people in my life today.

An Endless Stream of Days

The northern hemisphere tips towards the sun and the spring comes to this bit of the planet, and the people in the southern hemisphere begin their autumn. The changing weather pattern of the last few days seems confused and hasn't settled into place. Yesterday's sunshine turned into rain last night and now, although it is bright, there is a cold breeze blowing from Siberia, where the cold breezes usually come from. The cherry tree is holding tight on to half of its blossom; the rest rolls down the street to join great clumps of it that tremble in the breeze, like piles of foam from a burst washing machine. I roll over and cuddle into Peggy's arms and wait for the drizzle to pass. The clouds are thin and high and it won't rain for long, but looks like it could be on and off all day. I pull her back down into bed when she tries to get up, and she giggles and says, 'Aren't you going to work today?' as she snuggles down into my chest.

'It's raining,' I mumble. 'I'll go later.'

'No, it isn't, it's nearly stopped.' And as I turn to look through the window to see the clouds break apart, she jumps out of my grasp and runs off naked to the bathroom, laughing that she has escaped my clutches.

'Oh, bugger!' I grumble, both at the brightening weather and her escaping. I am feeling lazy and don't want to get up. I feel lazy more and more often recently, but as she is up and I'm feeling lonely in the bed on my own, I struggle out, go and give her a big cuddle from behind as she brushes her teeth. I have a shower, wrap a towel around myself and the day begins. A new day, another one in an endless stream of them.

It was not always like this. I was not always a good husband. When I was younger I behaved as if my life was about my happiness, and my struggle to find it. I was like a child. I lost a mother and a father when I was young, so I did not know any better. I thought Peggy's job was to give me love and make me feel safe and wanted, and when I didn't get those things I drifted away, life became dry and we argued, drifting further and further apart until we no longer argued and simply went past each other, naked and cold, brushing past stinging nettles. The kids had grown and left to live in their own places, and we were alone and looking at each other and wondering what our job was now.

One day I questioned if I should leave. I went hiking in the mountains, sleeping out for a few nights, watching the stars. Meditating. Peggy doesn't like the mountains and fears the unexpected drops that often appear as you round a corner. Heights, drops, falling, the unknown – I've made my peace with the unknown. As I clambered I pictured a life without her. I imagined all the details: being alone on those slopes and rocks, trying to build a new life, finding someone else perhaps or being solitary. Whatever I pictured I couldn't make it work, I couldn't see any happiness. I realised that my unhappiness was my own responsibility. If I couldn't make my life happy, at least I could try to make hers happy and be of some use in the world. When I got back I told her my thoughts and we clung together and cried like babies.

I love her again, and now again and then again. With each moment I become new, and each moment she too is new. So I begin each day from scratch, with forgiveness. There is no debt. I rip yesterday's pages from the account book and throw them into the wind. Forgiveness is the only answer. I forgive her for not being perfect, and she forgives me for not being perfect, and all that is left is love. We became heroes instead of cowards.

The heroism of love is in the clear knowledge that the object of our love will be lost, yet we commit to it anyway. I will not live to see triple figures. I am a working-class man and great age rarely happens to us, but that does not bother me – my life is my own, hand-made, and I have no desire for any other life. I have freedom. Peggy and I talked into the night about how long we might live, and for how many more years we might continue to be active and healthy. We talked about what it might be like to grow old together and what we could do to make it good. Peggy said that she was happy we were planning a future together. We made a plan called 'Ten Years of Fun'. I claim it was my idea, and she claims it was hers (it was mine really). A plan to have fun, within whatever our declining and increasing abilities are. So when some new task presents itself, or even some mundane regular task, one of us will ask, 'Will it be fun?' and if the answer is 'No', the other will say, 'Why are we doing it then?' And if the answer is 'Money', then the question is, 'Is it miserable money or happy money?' If we decide that it is miserable money, one of us will say, 'Well, I am not doing it then – we will just have to live off bloody lentils again.' My life now is about making her life happy.

Today I am pleased when the sky turns black again and begins to throw down fistfuls of fat drops. Although I'm dressed, I go to lie on my bed and read Seneca while Peggy potters around, and we have a lazy couple of hours and then I say, 'Let's go to the sea.' So we rattle to the sea in the van and wander around in raincoats, holding hands. We eat pizza, and I buy a new teapot and take photographs of the waves. On the way back I listen to a weather forecast on the radio and it says there is a heatwave coming. The weather forecast is the only news I listen to these days. I live the life of a happy idiot.

Fossils

Sunshine. I'm kneeling like a supplicant on my foam-rubber mat by the bed that goes around the pond. A pond skater stands on the water's skin, its four long feet make little dips, lenses that bend the light and shimmer; holding its front legs up as if in prayer, it looks into the depths, ready to grab with its legs and stab with its beak when it feels something moving below. Above the water a mass of tiny mote-like flying insects swirl and enjoy the warm air. This is a difficult little bed and badly designed. Having a circular wall around the pond means that there are marginal areas that get glancing light; the rear is the dark side of the moon and always in the shade, the front is always in direct sunshine. Being right next to the old leaky pond, the ground is always wet. Nothing ever succeeds in this stale soil other than weeds, and I have considered growing tall bog plants all around it – iris or something – so that the ones on the shady side peep above the wall and get some sunshine. Dandelions have seeded in it, and I am digging with a trowel to get all the way to the root of a particularly difficult one. I know this dandelion, I have had dealings with it before. Yes, I know how that sounds.

Last year I tried to dig it out and part of the deep taproot snapped off. I left it to return and haunt me, so now the taproot has grown back bigger and the top has forked into two separate flowers. I decide to dig all around it and go deep. This long root is as thin as hair at the bottom and tangled around something hard, so I dig deeper into the soil, making a great pile nearby on the bed. I pull out a rusted old tin that breaks into pieces as I tug at it; inside there is a clump of mud and, emerging from that, the bonnet of a toy car with traces of paint remaining in the rust. I pick some of the bits of mud away and slowly the toy emerges – a fossil of a plaything that still has its hard, but crumbly, once-white tyres on metal hubs with a rusty steel axle going through them. I pull out another decayed toy that looks like it was a delivery van made from steel, pressed and held together with little bent tabs; the remains of an aeroplane and a rocket, made of die-cast white metal, decayed and furry white with oxide. The aeroplane has its wings and stumps of propellers. The comic-book rocket is almond-shaped with a couple of red fins at the back, where there were three; the red, thickly painted nose is still bright.

They are quite lovely in their decayed state, and immediately I have images of a small boy in short grey trousers and a grey hand-knitted jumper sitting here by the fountain, seventy or eighty years ago, playing with the cars on the top of the wall and zooming the aeroplane over them, launching the rocket and making explosion sounds or pretending to have landed on the moon. Living his day as if it were his last, so neither fearing nor longing for the next – the essence of play. His mother calling him in to dinner, and the boy putting the toys in the tin and running off and forgetting about them. Or burying them, as if they were treasure. They have been here for many years. I brush away the loosest of the earth and line them up on the top of the wall to dry out. Maybe Miss Cashmere will come across them and recognise them. Perhaps they were hers or she had a brother, or they were here before she bought the house.

I wonder if it is a family house that she inherited. Suddenly the house has a history that I have never considered before. It has always just been Miss Cashmere's house. Perhaps she had a brother who died, and these things would remind her and so should remain buried; perhaps she buried them herself. And for a while I have very mixed feelings about the danger of unearthing a past that should be left to lie. But I fill in the hole, then finish weeding the bed and pack up my tools and leave the toys on the wall. The soil you take out of a hole is always bigger than the hole it came from. Digging into the past always pollutes the present. Writing this book, I scrape a little mud off the past and the feelings that are released have nothing to do with this moment, and I wonder why I do it.

I turn the corner and there is yellow all across the fields, and in my work I feel rewarded for the effort. In the smallest things there's poetry for me. I prune my own rose, which nobody else will see. Live my own days. A strong, warm wind sends cherry blossom spinning round and round me, as if I was the only bachelor left to wander home alone after the wedding. The may-white blossoms of hawthorn and blackthorn, frothy cow parsley, fern and hogweed, wild carrot, thistle, dogrose and bramble. I have only one desire in this world: to make it a little more pleasant – even the tiniest amount would help; the past rarely does.

Night Scents

Night-scented stock (*Matthiola longipetala*). She will not see it or smell it perhaps, and yet I have the seeds and so I'll sow them in a patch where someone might wander through a cloud of perfume in the dark. The flower is little to look at, but after dusk it sends a trail so rich and sweet to pollinating moths. Anybody passing by could not fail to notice it and turn and say, 'Where is that amazing scent coming from?' I am sowing it for the flower, the scent, the moths – unable and unwilling to disconnect this chain, unwilling to make a different choice.

Sweet woodruff (*Gallium odoratum*) that I planted years ago is covering the ground around the wild, self-seeded holly trees. It's green and pretty, has tiny flowers and long-petalled leaves, arranged like daisies around the wandering stems. The roses are starting to come out and, although I love to work with them, to prune and shape and feed and nurture them, they are not flowers that I have a passion for, pretty and varied as they are. They are overblown, hybridised and messed around with, to suit the vanities of man, and I see in them the ordinary and predictable lust to control and dominate. I prefer the weeds and self-seeded plants that grow where they will: the aquilegia and daisies, the poppies, the dandelions and foxgloves. Although I grow and pamper the roses, they are not for me; the plants that I grow for me are tucked away in corners, in my sweet / mad little plots, and often just where they landed.

The bluebells have faded, the tulips are flagging, the daffodils long gone. It's all azaleas and rhododendrons and iris. Blasts and pops of colour everywhere. The copper beech, dark and full with deckle-edged leaves, has tiny bronze and hairy seeds already – beech-masts. The chestnut candles, each a nine-inch pyramid of small pink-and-white flowers, are attracting flying insects to forage and fertilise and turn them into conkers. The wild garlic leaves have done their work of feeding their bulbs with sunlight and have collapsed, the white snowball flowers fading as the low sun sends its rays into the shady places where the garlic grows and dries them out, making patches of light and shade between the shrubs and trees. The ferns have unrolled and are looking young and hungry.

Miss Cashmere is coming down the path in a yellow summer dress, bare-legged and wearing sandals, carrying a hat in one hand, a newspaper and a pack of cigarettes and an orange disposable lighter in the other, heading for the summerhouse to read and looking sprightly, young and hungry. The warm weather brings out her colour, and she smiles.

'Good morning, Marc, and how are you?' she asks on this fine day.

'I'm very well, thank you, Dorothy. And how are you? You're looking well.'

'I'm good, I'm good,' she says. 'Just off to read my paper, though I'd rather not,' she says, as if she has no choice. And perhaps she doesn't have a choice; perhaps it's part of her make-up that she has to know what's going on in the world outside.

I'm happy and inwardly silent in this walled-in floating world of flowers and empty nonsense.

'It's looking lovely, Marc.'

'We do what we must, and luckily I really enjoy what I must do.'

'You are a lucky man indeed then, Marc.'

'I am, Dorothy, I really am,' I say.

She wanders down the acres of garden that she owns, behind her ancient family house, to sit and read in leisure, while I kneel and labour before her with my trowel. I am a very lucky man.

Burning Books

The twisted pink Xs of clematis wind up the holly tree. Broken propellers crashed into the thorns, with tendrils falling down. Big pink oriental poppies coming from their globes, and petite Welsh poppies alongside. The daffodils all gone. The bright-green stems turned brown, the seed heads snipped off and the leaves left to gather sun and feed the bulb. Tulips, too, are gone. I plant out the cosmos that are big and strong, with flowers budding, in their pots.

Miss Cashmere seems happier than I have seen her for a long time. She smiles and is relaxed and is back to prettying herself up. Something has been settled for her, and I wonder what it can be. There have been more people coming to the house – a carer in a pink uniform comes three times a week – and she has a new stick. A man in a suit has been around a couple of times. She left a pile of books in the greenhouse; no note, just a pile of books, so they stayed there for a while as I wondered why they were there. Books of poetry, a book on Japanese painting, a couple of gardening books; really ancient gardening books.

The pile grows, and eventually I catch her and she tells me that she is having a clear-out and anything I don't want, I can throw on the bonfire. Old novels start to appear, ancient novels perhaps from when she was a girl.

'I'm getting rid of lots of stuff,' she tells me. 'I'm tired of all this old tat and it has to go! Do you need any furniture?'

'I don't think so, Dorothy,' I say. 'But thank you.'

Through the windows I know that her house, downstairs at least, is full of antiques. I cannot stand collecting old stuff; collecting any stuff at all really, apart from books perhaps, and I am trying to cut down on them.

I finish work early. As the flowers come and nature takes over in the spring, my work dries up and things in the garden get on with their job of making more of themselves. The grass has to be cut, of course, and there are things to thin out and deadhead; a few shrubs that have finished flowering need trimming; but it isn't hard work and at this time of year, as it gets warm, I like to save a little work for tomorrow.

At home I move my chair outside to sit by the open back door with the sun beating down, and drink tea from my old brown mug that I bought from a potter who made it; it is rough and a perfect fit for my hand. There is cuckoo spit on the lavender, the sparrows sing in the late-afternoon distance, and Peggy sits inside. It is 7 p.m. and hot. We are silent and happy to be silent, and I think of the heat of the coming summer. I hear her pages turning. Heat from the oven fills the kitchen behind me with scents of cooking cheese and onions. A small honey bee seems lost, wandering quickly on the path; she seems to be trying to fly, but can't take off, the odd little skip forward and then resuming her fast walk, then another little skip, until she disappears into the plants and I wonder what, if anything, I could or should do about it. She returns, steps over an ant, who is very clearly and purposely on his way to or from somewhere, while the bee seems aimless, going round in circles.

Above, I count twelve vapour trails of aircraft crossing each other, drawing boxes, making woolly graph paper of a sky that is otherwise empty – the deep blue of a workman's shirt directly above, and fading to old worn and washed-out denim jeans towards the horizon. Then, as the sun falls behind the house, the temperature drops and I bring my chair back into the kitchen and wonder what I will do with my evening. I am not interested in my book, the television depresses me, but Peggy goes into the living room to turn it on and drops onto the sofa. Late-spring and summer evenings leave me cold and empty; they go on for ever, I am too tired to read or do anything, but not tired enough to sleep. I stay in the kitchen and watch through the window as the last rays turn the sky pink and the wood pigeon calls and the blackbird takes his post.

Sparrows flit between the apple tree and the lilac that I planted years ago, its old brown flowers rusting and orange in the falling sun that sinks behind a holly, piercing through the sharp leaves, making skinny, ghostly silhouettes of thick branches, as I squint my eyes into the sun and give myself a headache. Suddenly, as I refocus, I see the sky is filled with tiny flying insects as the low sun highlights them. Millions, like seeds as far as I can see. When I turn back into the room, it seems black until my sun-flashed eyes adapt. And, satisfied, I go through to sit by Peg and watch her TV programme until, as dark arrives, she lies across me, lays her head on my lap and denies that she is falling asleep, until she stops denying and her head suddenly goes heavier.

Sun!

The sky is bright and clear and looks as though it will last for ever. I work in shirt sleeves and short trousers, I cut the grass and feel energised. The warm weather brings a change in diet; stodge is off the menu, and green and bright colours are on, and I seem to work faster and get more done. Hibernation is over and the time of growth and production has returned.

The sun lasts for four days, and Miss Cashmere in her summer dress, happy now, wanders down each day to the summerhouse with her newspaper and cigarettes, her ginger cat following her to sit on her lap when she settles. Her routines are back to normal; she smiles and I smile and a few words are shared between us, our long-distance connection old and thin, yet pleasant enough. The summerhouse where she goes to read her newspaper holds a collection of bright things: a sparkling glass chandelier, table lamps with glass pendants, crystals hanging from nylon threads in the windows. Each corner of the pale-blue-painted room has fractured spectrums of colour moving restlessly across the walls and the ancient comfortable sofa, piercing the shady places. It is, to me, briefly attractive but the attraction wanes; there is no space for restful shade in that kaleido-scopic box. I need darkness to wrap me up, shadows to look into, mysteries that contrast with the clarity. The childlike coloured rainbows and sharp shards of white cut into and slash the tranquillity. There are no layers of shadow, just sparkles that take their mystery, peace and warmth away.

I am a creature of the shadows. I like places where light shades and fades away into many depths, and draws outlines and distorts and replicates shapes, and forms become something other – something richer that tells a simpler truth about the things they are made of. These shadows are the very essence of shape and form; without them there is no form and all is flatness, plastic and white. Without the shadows I am blind. Without the shadows, life is disposable.

The pile of yellowing, browning books in the greenhouse grows and pages have started to open, and the shiny trails of snails cross curling covers where they have eaten cotton and cardboard with their rasping tongues, and I think that burning them would make an awful mess because books don't burn; they just smoulder, then go out and leave a pile of sodden, half-burned slobber that has a throat-catching stink. I know: I tried it in the fireplace of a derelict house where I was sleeping years ago, to keep warm, and I ran out of matches trying to get the books going. I look at the dampened titles for treasure. The heat and damp shade of Seamus Heaney's *Death of a Naturalist* goes in my bag, and out of the cobwebby greenhouse that now smells of old books. Ginger cat under the bench watches, as if she knows of Heaney's lines where kittens are drowned as pests on the farm.

There is always a lull for me at this time. The plants are busy doing their thing, growing and pollinating and making seeds, and there is rain to water them; and nothing is dying and nothing needs protection, and the soil is good and fertile, and my busy work of spring slows as the mature growth of summer takes over. I slow down, take my time, sharpen my hoe with a stone, then hoe the beds, leaving the cut weeds on the top to dry in the sun, working slowly with sleeves rolled up, wearing a straw hat to keep my head from burning. Half bald, I shave it now like a monk. I like to feel the breeze and rain on it, like a rock in a stream. Acers are already making their little two-winged seeds, looking like bright-green flies in hanging clusters. Bright-blue ceanothus, rhododendrons and azaleas in vivid yellows and pinks and darker reds all vie to attract the bees and flying insects. I hoe herb-Robert (*Geranium robertianum*), our own sweet hardy geranium, also known as cranesbill, because of its long, pointed seed head. The tiny pink flowers sit on hairy and juicy red stems – a weed, a native wildflower used as a medicine for kidney stones. Their strong smell moves along with me as I cut them off.

On a rose leaf a two-spot ladybird has a male on her back; the only way to tell them apart is because the male is on top. They can stay together for eight or nine hours, and he will ejaculate a number of times. She will store 600 eggs' worth of sperm inside her, and when the weather is good she will fertilise them herself, about twenty at a time, and lay them in the middle of a colony of greenfly or other aphids. In a couple of weeks the eggs hatch and out crawls a larva, an imago, a wingless bug, dark grey or black with yellow spots along its length, with six short legs at the front end and a segmented tail. Many of the eggs are infertile and will provide food for the emerging hatchlings, to get them going before they tuck into the greenfly. As it grows, it sheds its skin about four times until it forms a pupa, hangs by its tail and metamorphoses into a ladybird, which struggles out in August and continues eating throughout September. If there has been plenty of food about, the ladybird may survive the winter, sheltered in cracks in fences and trees; if there has not been much food, she will desiccate and die in the drying wind.

Heart

Pushing my mower in the heat of midday, I begin to feel a pain in my chest, my pulse goes rapid and irregular, my breathing tightens and my energy falls down, down my chest and arms and through my feet back into the earth, and it is only through a massive effort of will that I remain upright and do not follow it. There are pains in my arms. Pins and needles in my hands. I release the dead-man's handle and the mower's engine chugs to a stop. Silence. I lean on the handles for a while – thirty seconds, three minutes, who knows – just waiting for my heartbeat to stabilise. I put two fingers to my neck to feel my pulse: the beat is rapid and very irregular. This is an old companion, back on the scene, having been away for a year or so. Medicated, I almost forget I have it, reminded only by the little bitter pills of poison I have to take each day.

I push the mower back to the shed. Drive home slowly, park, enter the house, go upstairs, strip and go to bed. Peggy comes down from her writing room in the attic, sees the familiar dark shadows under my eyes, my pale skin, and says, 'Fuck, do you want me to get you anything?'

I shake my head. She sits on the edge of the bed and holds my hand. Getting home, getting into bed, has taken every molecule of energy, every spark of electricity, and I am as limp as a rag. I lie and close my eyes, panting.

'Do you need to go to hospital?' she asks.

'Not yet,' I say, 'let's see how it goes.'

She lies down next to me and holds my hand for a while, she wants to get close; she looks worried, afraid, and I tell her not to worry, that it will pass, that we know what this is and how it goes. I feel like weeping. Problems with the heart have an emotional impact – ask anybody who has had a bypass.

My heart has some intermittent electrical problems. I had an ancient red Volvo like that once; it had to go. I had work done on my heart in hospital, but every now and then it still goes crazy for a while, and I pant while it hammers as if it wants to break out through my ribcage, like a demon out of its red-velvet hell, ripping through the wet ropes that tie it down and just take off, or rev faster and faster until it throws a con rod through the crankcase and spills its oil all over the road. The 'lub-dub, lub-dub, lub-dub' of a regular heartbeat has become a 'bum, bum-bum-bum-bang … bum … bang bum-bum-bum'. Its beat can go to 200 a minute; sometimes it skips several beats, sometimes it goes down and beats only thirty to forty times, as each slow minute creeps by. It quivers like a jelly and doesn't pump, and the blood can pool in eddies. Whirlpools where a clot may form. If this happens there will come the stroke, which will likely leave me mute and wheelchair-bound for the rest of my life. This is not the inevitable outcome, but is a common one. Rest helps. There are triggers that I avoid. Coffee, red wine, drunkenness. Yet sometimes I forget, throw caution to the wind and get drunk, especially if there is a party. I do like a party.

Maybug

listening to daybreak
there is no separate me
who is listening.

A dead maybug on the grass, they don't last long – like fireworks. I am rested, my heart is stable and peace reigns over me. A week has passed and I think about the half-cut lawn. I spend the days in my chair outside, wearing a Hawaiian-print shirt and a sarong, reading all the poems by Stevie Smith and smiling at her silly little drawings with their hint of menace. I am slow and happy.

I have seen death a few times; it holds no fear. I have seen the life drift out of people and animals until it failed to return. I have killed animals, watched them be killed by other animals, and I have been close to my own death. Close enough to accept it as a friend that teaches me how to live well, joyously and connected to the world around me as we change and look into each other.

Rain, No Rain

Breathing in and out
The rain comes and then it goes
All is just perfect.

Daybreak comes, and my jug of a head is empty and light to carry. Through the front window there is poppy-seed-sized rain making swirling shapes in the light breeze. A robin shelters under a willow. As I go to make tea, I see that at the back of my house there is no rain at all and the sparrows continue to fly. The shower passes. Breeze rattles the blind at the open window. Birdsong comes in on the damp air. I hear a neighbour open her back door.

The catalogue of spring bulbs arrives in the post. Hundreds of varieties for ordering now, to plant in the autumn to flower next spring. Lying next to me on the sofa as I open the mail there is a copy of Sylvia Plath's collection of poetry, *Ariel*; it was published in 1965, two years after she chose to end her unbearable world by suicide. Just for her, I decide to buy tulips, red and pink tulips. I fill in the form in the back of the catalogue. I decide to go to work, as the drizzle is so light; there is maintenance that I could do if the rain gets heavier: clean the mowers, empty the grass box that I left half-full last week.

*

The daisies are not yet open, fields of pearls straining against their tethers to blink open their eyes. The grey sky stays grey, and many of the little white faces do not bother to look. A dozen jackdaws stand on the grass, waiting for the moisture to bring up worms, but the earth around the stables is compacted through overuse and tree roots, and there are no worms. One by one the jackdaws flap away, slow and downcast, to find a better place to feed; they form a circle on the lawn, all facing east.

I busy myself with slowly tidying up the mess I left. My feeling of weakness is gone, but I am nervous about triggering it again, so I take it easy and focus on small, light jobs: sorting out pots and washing them, cleaning and sharpening tools. On juniper and yew, bright-green tender growth stands out against the dour darkness of old conifer. I pass under the massive copper beech, with its dripping wet trunk, whose roots spread out above the dry soil; ferns grow between them, and ponds form in cups and hollows, and ivy wanders in the shade. Where bats roost and, in the evening, fly about my head as I look up, I try to see them sweep by so fast. The owl watches and hoots as I pack up to leave. It is late. I have been catching up. There are beech-masts falling already from the heavy branches. The trunk is green on the back, with orange and pale-blue-grey patches of lichen at the front where the dappled sun can sometimes reach.

Jackdaws, desultory, stand in the drizzle and barely bother to get out of my way as I pass by acers and their ripening seeds. The candlesticks of conker flowers have gone, but in the shade a few magnolia flowers hang on, and rhododendrons and azaleas that have been blooming for weeks are now fading, turning from pink and orange and yellow to gorgeous fading rust. This ever-changing here-and-now absorbs all of my attention. I will not waste a moment of my life imagining what could and could not become, re-imagining what has or may have been.

June

A Dumb Labourer
Visits

Unwrapping another day. Should I look for fresh clothes or grab last night's damp ones? Either way feels good; a new day in old clothes that feel as familiar as last night. I open the blind and take the peel off the juicy day. The seagulls are laughing and screaming like a pub of mixed drunks. A whirlpool of them, slowly cruising above as I get ready for work. I guess they are celebrating the birth of a new one, the cracking of an egg. It is 1 June and it feels like the beginning of summer. The gulls turn about above the land; there is no sea to float on, and I wonder if these waste-tip gulls have ever had the pleasure of bobbing up and down on a moving wave, catching a fish instead of a half-eaten kebab.

Under the beech tree, which last year hummed with flies and bees all summer long, the growing seedlings are scattered in random, Zen-like patterns that no novice monk could ever make. They swallow up the sunshine, the red and blue reflecting back the green bits of the spectrum they do not need. Feeding on the innocent, my spinning petrol-driven jaws mow them off, along with dandelions and buttercups and vetch that arrived soon after Mr Cashmere died. The mower and I promenade up and down, past peonies, roses, foxgloves and leafy hydrangeas just starting to open their flower buds. Tucked in a corner, trying hard not to impress, a group of yellow Welsh poppies (*Meconopsis cambrica*) mixed in with aquilegia cheers my heart with their simplicity. A single blue Himalayan poppy (*Meconopsis betonicifolia*) with strappy, dusty blue-green leaves, which comes each year and flowers on the shady side of the beech hedge, makes me stop, let go of the levers and, as the machine runs down into silence, the poppy's impossible water-blueness seems to go on further than the sky. It is in our nature to enjoy a rare thing. When we find such a thing, we want to make more of them until they are commonplace and we don't find them precious any more. I leave that one blue poppy, I won't make it common. I will not propagate it; it will be rare and beautiful and, when it dies, there won't be another.

When everything else is gone, the love and hate and distracting toys are gone, the fictions and lies are gone; when the gardening is gone even, and the poetry fails to satisfy, all that is left is nature. I try to remember Walt Whitman's famous quote about this, but perhaps the distance between us has become too great. So forgetting will have to do until I get home. Perhaps I'll bother to look it up; perhaps I'll not remember and will simply steal it as my own.

Miss Cashmere wanders down, carrying her usual bundle: newspaper, cigarettes, lighter.

'You look thoughtful, Marc, are you okay?'

'I am absolutely fine, Dorothy. I'm just enjoying the poppies.' I feel guilty for a flash that she has caught me idling yet again, but I let the feeling pass by; it's part of my job to look, but the dumb, horned labourer I was bred to be, who thinks that I should labour all my days, comes out – low and weak – from his dark place. You never get rid of anything. All you can do is add to it, build strata and hope to bury the beast far down through coiling layers of learning. I let him wander off back into his maze and I say, 'The blue is stunning, isn't it?'

And we stand together, looking at the single Himalayan poppy and, below, another bud not yet open, a dull, matt blue-grey-green ball with traces of heart-breaking blue peeping out; we look for three seconds, or two minutes or five. 'I do love your garden,' I say softly through my trance, so quiet it almost feels like I said 'I love you'. 'It is a special place,' I say. And for the shortest moment I wish that it was mine and I didn't have to work, but could come and look, lie naked on the grass that I have only ever kneeled or trod upon, or sit on the edge of the pond with my trousers rolled up and my bare feet in the water with the fish – my big toe the right size for their gaping mouths. I let the fantasy pass by and let myself be happy that it isn't my garden. I don't want ownership – not of anything really. Ownership goes both ways. The owned owns the owner. The more one has, the more one fears losing what one has. I do not want the misery of stuff. My cushion next to my own back door, and my cherry tree and Welsh poppies, my ever-growing collection of poetry books, my wife and children – they own me and that is enough.

My grandfather used to say that I changed like the wind, and I do. I'd rather be a vane, a cock, a kite tugging at the string to go this way – then this way – then this, and not force myself to anchor to a patch of ground or a belief, while the wind moved around me. It is a skill I have practised for years. When the idiocy comes in, or a sadness, or an anger or judgemental thought, or an insubstantial plan or memory comes over me, I have learned to recognise it right away and choose to switch it off or at least to understand. When happiness or joy comes in, I recognise that, too, and let it flow. These feelings are not real things. They have no substance; they are hypnotic states created by a mind that would continue, if I let them – ghosts that can be stopped dead by another idea. I can stand in a field for hours, or work all day and think about nothing and simply be ruffled by the breeze like a common roadside campion, where each moment can be a lifetime and I can live for a thousand years.

Hungry? Eat. Tired? Rest.

If my eye is attracted, I look.

I let myself fall into the otherworldly blueness of the poppy, with its hairy yellow centre, and all that remains is nature. The buzz of insects, the breeze.

Then Miss Cashmere says, 'Are you happy, Marc?' We never, ever talk about personal things. I look at her and she looks at me intently, her pale-blue eyes set deep in their old, wrinkled sockets like bright glass, distant planets, poppies shining with youth.

She has a serious expression and I say, 'Yes, Dorothy, I am deeply happy. My life is good and balanced. How about you? Are you happy?'

She obviously asked me because she wanted to talk about her own happiness. The words she was preparing unroll from her like silk from a cocoon. 'I do not know,' she says. 'I'm not sure what that is any more – happiness. I used to be happy, when I was young. I'm not afraid, I have nothing to worry about; my children seem happy and settled in their relationships and their work, there is nothing I need. And yet I'm not sure.' I don't respond. 'I don't feel anything,' she said.

'What do you want to feel?' I ask. 'Are you content?'

'Yes, you know, Marc, I think I am content.'

I am becoming confused by words and try quickly to develop a scale of happiness in my head, and place 'happy' on it somewhere and 'sadness' on it somewhere else, and 'contentment' somewhere and 'joy' somewhere, and 'bored' somewhere else. And what about 'wonder'? Does 'wonder' go somewhere on this linear measure? But in the end the word-soup feels like ideas that should not be on a scale at all, a distraction. Nothing built on the shifting sands of words is to be wholly trusted, although they have the appearance of trustworthiness, which makes them even less trustworthy.

'What do you want to feel?' I ask again.

And she ponders for a while, then says, in a dreamy, thoughtful way, 'Perhaps I am bored.'

She is wearing a long summer dress, her pale ankles are bare and veined. On her feet she has flat yellow canvas slip-on shoes that are wet from the grass.

'You should take off your shoes and walk barefoot in the grass.'

'What for? Why?'

'Like a child,' I say. 'It could be fun.' She laughs as if embarrassed. 'Walk in the wet grass. Leave your shoes here. I'll take them up to the house, so they will be there for you when you get back.'

'Oh, I don't think I could do that,' she says, but she doesn't move.

I think I can see that she wants to, that she likes the idea, but is concerned about how she might appear. Or perhaps she really doesn't want to take her shoes off, but also doesn't want to refuse and thereby leave us both feeling uncomfortable. I am not sure now, and I'm beginning to wish I hadn't said anything. I don't want to embarrass her and, while I am trying to think of a way out that would leave us both feeling comfortable, she puts her hand on my shoulder, touches me – I don't think we have ever physically touched each other before – then uses a foot to take off first one shoe and then the other, and she leaves them there, in an act so impossibly generous that it pricks my eye.

Very dignified, she crosses the grass to the summer-house and grabs her skirt to keep it off the wet grass.

Watching her leave like that, I don't want her to leave, yet I enjoy watching her leaving. I pick up her warm shoes, worn, with dirty broken insoles, and I think of her bravery and feel humble. I take them to her house and leave them on the top of the wall, so that she doesn't have to bend to pick them up. I've seen a part of her that I've never seen before. I am worried that if she steps on something sharp or is stung, it will be my fault.

I'm on my knees, weeding, and the plants and the soil and I seem to flow into and out of each other. We are brothers and sisters, fathers and daughters. The scent of soil fills my nose, vibrant with small life. A bee flies from foxglove to foxglove, lands on the bottom lip of the flower, crawls inside the flower that fits her like it was made for her, with heavy bags of pollen on her thighs, then leaves and wanders, laden, drunkenly into the next one. My hands are in the earth with the worms and bugs and fungus, and the slimes and roots, and I pull out pink speedwell and juicy red-stemmed herb-Robert from around the foxgloves and lupins. With a ladybird on my bare arm, I feel something else land on my warm back, which I ignore: a shield bug, a leaf or a twig or bird droppings; a hoverfly inches from my face watches me intently and I watch it back.

Later I see her returning slowly to the house. She passes by, smiling, concentrating hard on where to put her feet. She leaves wet grassy footprints on the warm stone steps. They evaporate quickly and she is gone, and I am embarrassed and then liberated. I broke the rules and enjoyed it, as usual. Breaking the rules has always given me pleasure, often followed by pain. How far away from the plants and birds she had seemed. All she had to do was go quiet and let them in; they were knocking, knocking on her door and she couldn't hear them.

A New Path

A dead blackbird, folded wings and perfect on the path, no maggots yet. His song over. A summer rain that makes the day dark and murky, makes a liar of the thermometer, feeling colder than it says, and dismal. I work and thousands of years pass. I slowly carry my bucket of weeds to the compost bin, piles of cranesbill and dock.

I do not mourn the dead blackbird, or the grass I cut or the weeds I pull, although they are my sisters and brothers. For lunch I have boiled eggs and I do not mourn them, either. I do not mourn my mother, lost when I was a teenager, or my father or my grandparents, or my childhood or my homes, or anything past. They are all equal, and the joy is in their having had an existence at all. There is nothing to be done about them, other than decide how to think about them; they are vague and lost in the fog.

She wants me to make a path down to the summerhouse using the pile of old limestone flagstones behind the stables that have been there for decades. Thinking about the 'desired path' – the route she would take in dry weather anyway – I lay out a row of slabs along a route, avoiding steep slopes or steps and following the contours; it curves one way, then slightly back in the other in a stretched-out serpentine shape that curves gently down. I leave the slabs there, to see what she thinks of the route the next time she passes.

Quaking aspen catkins hang red and hairy from the branches. Acers' moth-like seeds cluster above my head, and spiny red beech-masts litter the leaf mould underfoot. High above in the massive limes, little groups of balled seeds – sometimes six, sometimes eight – hang on threads from a single dangling wing. And where, three short weeks ago, there were bear-like masses of white-and-pink flowers on the horse-chestnut trees, they are gone and tiny conkers, the size of my thumb, bright green and spiky, fatten and ripen in the sun.

Behind the trees and by the stables is the little 'Ty Bach', the outside toilet, which is mine alone to use, though when I need to piss I do it on the compost. Next to it three water butts, and a grooved and wobbly sandstone sharpening wheel with a rusty crank-handle on a weather-beaten mossy wooden stand, still sturdy after generations of use. A stump where I split logs and sit to sharpen blades, axe, scythe, knife, trowel, shears, secateurs, half-moon edging spade, trowels, hoes.

I work around the buildings with the brush-cutter. It screams and makes smoke, a violent, senseless thing that slashes back the grasses and native wildflowers. A 'weed' is a word that tidy-minded people use for plants they do not want. Anyone who loves the earth knows that tidy-mindedness is death for nature. I am a wildflower, an untidy weed. The scent of petrol, engine fumes, hot oil and blended greenery fill the air, and behind me the meadow is flourishing. The machine is violent and stupid. The violent and stupid nearly always win; it's why they are created: to fight and win for their owner's gain.

She comes by again. She is brisk and well drawn in dark, slim trousers, white blouse, dark short jacket, flat black canvas shoes, white hair pulled back. Carrying her newspaper and cigarettes. She seems upright and businesslike today. Maybe she has good days and bad ones.

She is more talkative than she used to be, or is it my imagination? Perhaps we are bumping into each other more as the weather improves, as we are both in the garden. I see her coming, but pretend that I don't. I would rather get on with my work. I find human interaction awkward sometimes, and I have a lot to do, so I keep my ear defenders on and continue thrashing back the grass around the bricks, then work my way along the boundaries. I move slowly; this machine is vicious and there are places where there's wildlife that I won't use it, but instead resort to slower, old-fashioned methods, but here along the bottom of the walls and fence-line it is hard to cut in any other way. I keep the grass low, and I can see if there are toads or frogs or field mice in my way and take my time. There rarely are; they all head for the meadow and the marshy pond at its edge, which is full of tadpoles and newts right now; above it, hovering midges and flies of all kinds.

She passes and I ease myself round, hoping to appear as if I haven't seen her. But she is clever and has probably realised that I would rather not be disturbed. Later, as I rest and eat my cheese and apple, she comes by. She is carrying another book.

I want to harmonise with people, but do not know how to do it. I ask her about her trip to Japan and, in doing so, realise that she has never actually told me that she went to Japan. I just made an assumption. She tells me that it was rainy and she spent time with old friends. I asked her how come she had friends in Japan, and she starts to tell me about work colleagues that she has known for years, but then drifts away from the subject. 'You must have travelled an awful lot for your work,' I say – although she has never said what her work was, and already seems slightly resistant to my line of questions and tries to divert me – but I plough on: 'Where was the most exciting place you have been?'

'Russia,' she says, 'most definitely Russia. St Petersburg is wonderful, and such parties! Anyway, I must not hold you up any longer. The path looks good,' she goes on. 'Are you going to make it level or something?'

'Yes. I put the stones there to see if you were happy with the shape. I can always change it.'

'No, it's lovely,' she said.

'Good. I'll dig the stones in, so that they are level with the grass.'

'Lovely. I have brought this for you,' she says, handing me the book. 'I thought you might like it.' The book is a hardback, old and smelling of house, with a mustard-yellow cover with an oval picture of a black-and-white engraving of a serious young man, curly-haired and high-browed. I recognise him at once. The book is called *CLARE* in capital letters. I open it at random and read on page 53:

'*What Is Life?*'

And what is Life? An hour-glass on the run
A mist retreating from the morning sun
A busy, bustling, still-repeated dream
Its length? A minute's pause, a moment's thought
And Happiness? A bubble on the stream
That in the act of seizing shrinks to nought.

I do not know what to say to her. Suddenly, despite her distance, we are not so different. Even if she was a spy.

'Are you sure?'

'Yes, I'm sure. I'm pleased you like it. My house is full of old stuff I don't need any more, and I thought you would like this. Do you know John Clare?'

'I do,' I say. And wonder how she would possibly know anything about what I might like. Perhaps she and I begin to harmonise. We are all sung by the same earth, and if we spend enough time together quietly, harmony becomes inevitable.

As a child I rarely had gifts; at Christmas there might be a book, coloured felt-tip pens, a plastic recorder once, a transistor radio when I was fourteen. But the gift of this book – a deliberate gift, with the thought of giving it to me on purpose, and with no other reason than that I might like it – is a special thing. As I grow older, as we grow older, something passes between us. She sees that I am moved and seems confused; her expression changes about seven times in as many seconds and, oddly, we turn away from each other and she wanders off, saying, 'There we are then.'

And I, left standing there holding this book, feel a wash of grief that I do not try to understand.

Cold Returns

The temperature has dropped again, and it seems that the brief summer that was May is over. But it is still spring. There have been showers for days. In front of me now, a couple of yards away, a sparrow with her fledgling. She picks greenfly off the roses and feeds them to her chick, who opens his beak and takes them; the chick copies her, pretends to forage, pecks at the ground, but doesn't know what to do, so the mother keeps feeding him. I'm calling them mother and son, but I have no idea what their sexes are. So I'll change it to father and daughter and see how that works for me ... It's fine, and the simple binary gender change reminds me of the months and years I spent looking after my own daughter when she was tiny. Feeding her off a little plastic spoon from a heavy ceramic bowl that I bought, which had pictures of Peter Rabbit printed on it. Wrestling with her when she was bigger, teaching her to be confident, independent and easy at using her muscles and her body. She is so confident, now she is a grown woman, that I am a little scared of her. I miss her.

I don't mind growing older; in fact I quite like it. I am happy and content, I love my wife and enjoy spending my days with her when I can. Peggy and I are alone together again, like when we first met. We are free, and we don't have to go home at night. I think about my children, my son and daughter whom I see from time to time, and who have their own lives and seem fulfilled: working, earning, building relationships, doing all the things we ordinary humans strive to do. If they land back in the nest, I will be here to throw them up again; if they don't land, I will be here anyway. When I think about them, my throat closes and my chest tightens and my eyes prickle, and I miss them and want to hold them both again; want to be their father again, and for them to be little, so that I can play with them again and feel their warmth, hold their tiny feet. I have to learn to be independent of them.

It is not yet summer – just a few days to go. There are goosebumps on my bare arms, raising the hairs to trap a layer of warmth against my skin. There is some rain, not much warmth, but enough to let me roll up my sleeves. I hear a flock of sparrows; they are insistent now, here every day. I smell the marshy area by the meadow moving as the small drops fall, making splashes, crowns of water making kings and queens of sunken newts and frogs who lie in sucky footprint mud-holes.

Solstice

A blistering heat that makes the skin on my head glow red. I set the new path into the turf, cutting around each stone with the half-moon edging knife, lifting the green shadow of turf and dropping the stone into the hole. I pile the turfs upside down and make a little wall in front of one of the compost heaps. The sun beats on my back on this first day of summer, the longest day of the year, and I'm lifting and flipping heavy limestone pavers. According to the thermometer on the greenhouse, it is twenty-five degrees outside in the shade.

After lunch on the second day the path is finished and level, and none of the stones rock when I step on them. With a piece of string and a wooden stake I mark out a circle, perhaps six feet across, by the tightest arc of the path and dig out the turf. I pile the turf up with compost from the heap and, until the light begins to fade, I plant the white cosmos that I grew from seed in the new circular bed, for her to find when she wanders down. Clean and pure honey-stone under her feet, white flowers at the height of her hand, and green grass all around. It glows in the dark as I walk away.

The massed white flowers, with their strong balsam scent, on the cotoneaster tree are attracting millions of bees. In the autumn, thousands of berries will fall and stain the ground red. Because of the rain and the sun, the hydrangeas are flowering. Big mopheads in pink and white in beds by the house, and white drooping heads of flowers hanging off the house itself, where not long ago I was up a ladder pruning. By the summerhouse further down, the acid soil has turned the flower from pink to blue.

In Your Garden

I head to my bookshelves in a faded mood, and choose a book that is torn and faded to match it. Well loved and full of ancient *wabi-sabi*, that bitter sweet state of cherished decay, hiding in the shadows between the shiny and massive two-volume *A–Z Encyclopaedia of Garden Plants* and the equally glossy *Mushrooms*. My books are in no kind of order, just slotted in anywhere. Human hands have opened these pages since before I was born. Their skin cells and DNA are probably in the rough weave of the pulpy yellow-browning, ragged-edged leaves.

The book is *In Your Garden* by Vita Sackville-West. I look at what I should be doing today. She suggests that the iris are at their most lovely right now (which they are); she compares the falls – the drooping lower petals of the iris flower – with velvet and wine, and says they are the easiest plants to grow, and asks only that they are dug up and divided every three years to increase them, or to give them away. We have nobody to give them to and there are masses of them; the rhizomes have become straggly because I haven't dug them up for a few years – they are leaving their bed and wandering onto the lawn, and I feel neglectful. Vita also talks this month about taking rose cuttings. Having done this many years ago, we have a good bed of pink roses, all taken from the same few plants, which are looking great right now, but probably need deadheading. I put the book in my lunch bag and drive to work.

I trim a few iris rhizomes back with my secateurs and pot them up in the greenhouse, cutting their scrappy leaves back with shears so that they put their energy into developing roots instead of maintaining leaves. It was what I was taught many years ago. I do it out of habit. We do many things out of habit, and I remember a story I was told about a woman who always cut her Christmas turkey into pieces before putting it in the oven. After some years, her husband asked her why she was doing that. She told him that her mother had taught her to do it that way. She decided to ask her mother what the purpose was, and her mother told her that when her daughter was young, they only had a tiny oven and the only way she could get the turkey in the oven was by cutting it up. I am the same with gardening. I do it the way I was taught; my methods may be rubbish, but they work for me.

Then, as Vita suggested, I wander down to the roses and, not having brought a trug with me, I cut off the faded flowers and stuff my pockets with the immature seed heads. It feels good to let somebody else make the decisions for a while. It is summer, I am in shorts and a straw hat, feeling absent-minded and mooching about.

A Round of Applause

I wake to the applause of rain and wonder for a moment if I'm lying on my side or sitting upright against a wall. It would not be the first time I had woken in such a situation. I am horizontal with tattered pillows, the pillowcase creased and folded as if it has tried to crawl off in the night; it has made a matching fold in my cheek. I am a horned beast, deeply, gnawingly hungry, but I'm not sure what for; it is not food, sex, freedom, love, companionship, for I have these things. I do not know what I long for – I want more. It twists my gut, so I make myself go calm and let the feeling pass. A dream perhaps, passed through and forgotten, left a residue.

In our room the darkness is so vast that I can look hard and see nothing at all in any direction. I send my hand like a shark through the endless bed to touch her, maybe rest it on her hip or in the crook of her knee or, best of all, to hold the tight bundle of her hair, but it reaches instead the resistance of bundled sheets and blankets that she has rolled round herself like a larva, and I cannot push my way through without waking her. She is on dry land and I am at sea.

I fade away again. Some small time later, it seems, the dawn has broken through. Sleep hurries away like a mother on the scent of a bargain and I try to chase her, craving her warmth, but I am too small, nobody's flower. I try to turn away from the day, but that won't do. That will never do. It comes through anyway, however hard I try to block it out.

My breath's tide flows in and out, noisily. My oxygen level is low. My nose has been a poor organ ever since a girlfriend broke it, by sitting on my back and smashing my head into the floor during a play-fight. We were eighteen. She laughed so hard at her feeling of power when she saw the blood. I laughed, too, and gasped and rolled her off and kissed her hard to make her face bloody as well. I breathe through my mouth and disgust myself. I fear choking to death at night, like my mother, drowning on my own body fluids. I want to hang in a pool, perfectly still and corpse-like, and feel fresh cold water flowing through my blocked airways, rinsing them clean. I want to drown in clean. I have never in my life felt clean. The membranes of both nostrils are swollen, so nothing passes through this morning. The human nose has a cycle – one nostril flows faster than the other – and they are designed to alternate through the day. The membranes swell to narrow the opening and slow the air flow, so that we can experience a wider variety of scents, larger molecules travelling slower than small ones. Sometimes, when 'both channels' open, a new and precious world reveals itself to me.

Guiltless, I watch the Sunday sun rise. The dark fades to church bells, with no Pavlovian response from this heathen. Dawn arrives wet, but I don't have to move. From jaw to heel, my body hurts. I lie and listen to the beats and count them. Cars drive past twenty-eight seconds apart.

My heart beats fifty-four times a minute, but it is regular and for that I am happy. I eat magnesium pills to keep it in order. I'll burn up like a flare when my time comes. Peggy, in the bed beside me, is breathing once every second. I am trying to stop my mind thinking about me, so I think about her.

Peggy is a good sleeper. She is my woman and she curls, still sleeping. Mimi the cat curls, still sleeping, too. Two curled commas sleeping. I'm an exclamation mark banging around. We slowly navigate the route to wakefulness by touch. I find her hand and hold it. Peggy is quiet next to me, as if she knows I am elsewhere. I bring her tea in bed when I hear her wake, and the world seems good.

We chat about the day to come, and how well we slept and what she dreamed about, and then she gets up and goes for a bath. Under the sheets, I can hear her moving about and life feels precious. I hear her voice. She is laughing, talking to her mother on the phone about our son and his stupid cat. I go to find a book, my feet flattened, turned up at the edges, cold and splayed, on red clay tiles baked in a furnace – distant from their heat, like a dying sun, a black hole sucking at my vestigial warmth, freezing.

I take T. S. Eliot back to bed, and the rain falls in its own puddles and I am a child safe in my own world. Peggy, wrapped in a towel with her hair in a swelling turban, brings me coffee; she is brightly pink after her bath. The cat comes and lies on my feet and the world is perfect. My mug is hand-made, glazed on the lip and rough on the curved body. It fits my hand, it has texture and it feels precious and strong and brittle like a hollow stone, a relic, the bones of a life that somebody will have to deal with when I am gone. A brief awareness of this fragile life and, despite the nuttiness of the people who are pushing the Earth and human society towards destruction, it feels good to be alive and to be connected to the few things I have, and to the few people in my life.

In the distant future I can see the tunnel at the end of the light, but the life we have to live now, before the tunnel, seems brighter than ever. I pull on layers of wool and add logs to the still-warm black stove, and on my knees at the open iron door I fill my lungs with burned and dried wood-scented air and blow the fire back to life. The remains of a tree I cut down last year – apple – sits warming by the stove. I pile it in and check the damper. The logs are very dry and will burn slowly all day long.

We have made a place where we can turn into each other. Where the wind sucks the sparks and makes the chimney whistle. She cooks breakfast, I read, sometimes aloud. She cannot hear, but I read to her anyway. I read poetry, T. S. Eliot: 'I grow old . . . I grow old . . . / I shall wear the bottom of my trousers rolled'. I look over at my brown corduroy work trousers and their turned-up cuffs, hanging on the back of the door. A hum or roar outside and very distant, but we never find out where it comes from or what makes it. If I listen in this corner of the house, it is nearly always there. A machine somewhere or a transformer that resonates with this small place where I sit and read. I think

I should move my chair, but don't, because it looks good where it is and the light is good for reading and I can tune out the sound. I'm behind glass and watching the clouds and the crows. Peggy is watching, too, in a yellow dress. I love her. I love our little life. The daily tasks of washing pots and sweeping up, of pulling weeds and chopping vegetables: these are the things that anchor me and make me whole. It is Sunday. I do not go to work on Sundays.

I stand on the kitchen step, look over my own little neglected garden and I am spotted with rain. Pearled, and remembering the first time I wanted her. That was thirty years ago, when there was too much world to face. Now the world that remains is nowhere near enough.

I sharpen my pencil. She taught me how to write again, after I had spent so many years trying not to write, trying to do something more sensible. She told me to write and not to worry about what I am writing; to write as if nobody is ever going to read it; to write as if my words were the wintersweet blossoms that nobody would ever see. I sit for a while with my notebook and write a poem that nobody will read, about how the crow teaches me to let go, by abandoning himself to the wind and letting it take him where he wanted to go anyway. It isn't very good, but it satisfies me for the moment. It will be written again many times. It's more or less the same thing I always write.

Aphids

Blistering sun, then storms come. It's hard to work in either, so I shelter and watch flocks of sparrows swishing from tree to tree singing, a rush of plucking from a dozen mandolins. In the roses they are eating aphids; they leave the ladybirds, whose red colouring tells the sparrows that they are not for tasting. The ladybirds also eat aphids. By growing roses, I am nature's way of providing greenfly for the sparrows and ladybirds and ants. There is no shortage of aphids and they reproduce at a flock-sustaining rate. Rose greenfly generally live on young shoots. They can kill small plants by sucking them dry, and carry viruses from plant to plant that can make stems twist themselves into corkscrew shapes, or leaves develop mosaic patterns or curl and die.

The greenfly is a very curious animal. Squidgy and semi-transparent, they pass their winter as eggs, which hatch in the spring – all as females, which pump out live babies every fourteen days by parthenogenesis until they are surrounded by their tiny clones; there are no male greenfly at this point. The young emerge tail-first, and commence feeding on plant sap pretty much as soon as they find a gap in the crowd. From their beak they drive a sharp stylet down into the plant, and the pressure of sap in the plant loads up the aphid, inflates it with more food than it can digest properly. Overflowing from its anus, this drips as sticky sweet honeydew that attracts ants, fungus and mould. One individual 'stem mother' can give rise to millions, as they become sexually mature very quickly. Soon the mother is overcrowded, surrounded by her young, who don't travel very far. Some of the new females are then born with wings and fly off to other plants – one brood having wings, the next brood having no wings – alternating until the early autumn, when the various stem mothers start to lay males among the egg-laying females. They mate and the females lay eggs that survive over the winter, and the whole cycle starts again the following spring.

Some ants have a delicious relationship with the aphids. They farm the aphids and look after them, often transporting them to new food plants at the appropriate stages of the aphid lifecycle, sheltering the eggs in their underground nests during the winter, frequently running massive farms that can spread across the root systems of several trees. Ants milk the aphids for their honeydew, stroking their soft bodies with their front legs; the ants then use the honeydew to feed themselves and their brothers and sisters, and some species will protect the aphids quite aggressively from predators, spraying formic acid at them. The manna that came from heaven in Exodus is widely believed to be the honeydew secreted by an aphid – a scale insect that lives on tamarisk trees in Sinai. Aphids are also eaten by lacewing and hoverfly larvae, earwigs and beetles. I do not spray the aphids on my roses, although in the past I have lost whole crops of broad beans to them. I am nurturing sparrows and ladybirds, beetles, ants and underground fungus instead, all of which rely on the greenfly.

July

Stoics

I feel as if I have not yet been coloured in by the day. It has no plans for me, and neither do I for it. The feeling is good. I open my beak to squeak, but there's no sound to hear. Nothing to say.

On the roof across the road there are two grey and fluffy baby seagulls. Around them seven magpies hide behind chimneypots, waiting for the parents to look the other way, as if they plan to attack the chicks. Summer is here, it has been hot for a week; the red alcohol in the thermometer rises a little higher each day. I am watching swifts make whirlpools in the sky and surly, swarming jackdaws protect their young, who remain in their nests in the chimneys.

I am reading of the Stoic philosophers, Epictetus, Marcus Aurelius and Seneca. Epictetus was a slave who encountered one adversity after another – crippled, some say, by his master. He developed a philosophy that influenced the emperor Marcus Aurelius, who wrote extensively about it. Epictetus advocated unconditional surrender to the course of nature; he suggested that we accept and love how things are, and do not harbour desire for things we cannot have. I remember my time living rough and know this feeling well. Desire is the cause of all suffering, say Stoics and Buddhists alike.

Raindrops are falling from a hot sky that a moment ago was clear, but now is grey. As they hit the warm bricks they evaporate, and I think briefly, as I am in my seventh decade: if I were rain, would I be still falling, or would I have started evaporating yet? What would it feel like to be born and, drawn by gravity, live so short and quickly? Perhaps that is how we live anyway.

Drops fall onto the leaves, roll down and drip off into the earth to the roots and further down. A constant wind-blown hiss almost buries the background sounds of larger drops landing in water and smaller ones blown against the wall and fence, of the stream running in gutters, trickling and gurgling down drains. In the distance the rumble of vehicles, as people go about other temporary things. Two magpies cracking over crusts. High above in the grey, a seagull – white as could be – circles. Peggy reads and breathes and I make tea. These everyday things are fine things. How fine to watch the white cloud drift on by and listen to its children, cool air and birdsong.

I take Epictetus to heart and believe that every day could be my last on Earth. 'You could leave life right now – let that determine what you do, say and think,' said the emperor Marcus Aurelius. Today, after the rain has passed, I cut the yew hedge. I have a choice to use the big petrol hedge-cutter and do it in a morning or use hand-shears and take a day. I choose the hand-shears and have a pleasant and peaceful, slow day, smelling the air and the green leaves instead of petrol fumes, hearing the birds sing and the clipping of the shears in this, my childlike music box.

Wabi-sabi

Here is deep, warm summer and most of the work is done. The rose bed is filled with colour, and I seem to have made it look boring. The shadows too defined. There is a child's simplicity, a body with perfect features, superficial joyful colour splashed around, like the lid of a biscuit tin, with no imperfection there to catch and hold my eye. I have over-tidied and created the impossibility of weightlessness, when I prefer the devastating thump and squeak as the dancer lands. The little puff of chalk that shows his or her weight gives me a rush.

That skinhead beast Van Gogh's sunflowers bring forth in my mind's eye his self-portrait with a bandage covering his partially severed ear, a bloodstain showing through it, although in the painting there is no stain there. A child is adorable because their physical perfection is seasoned by an ungainly walk, a funny laugh, crazy hair, incongruous glasses or an overwhelming interest in a simple beetle. I prefer a rusted car, a chipped mug, a twisted bow tie, a wrinkled suit, a creaking gate. I strive for perfection for my customers' sake and then, for mine, I add in little 'flaws', a bit of the Earth's salt to make it less saccharine, less fake, less chemical and sweet, to make the merely pretty into something moving. As I pass each rose I brush it just a little, with my hip, deliberately accidentally make some petals fall, white and browning on the neat grass, and now it truly is complete.

These long, hot afternoons are relieved only by the early mist, the dew on the meadow and high, circling, ever-trembling swifts; the lower crowds of jackdaws fall on my thrown crusts and cheese rinds and carry them off to their dens. The young who, although able, seem unwilling to fly as if they can't be bothered, forage around, looking for dropped morsels, and try to get their fed-up parents to feed them. On the roof ridge fat pigeons sit in rows, a wing's length apart, occasionally take off, fly in some sort of circle as if testing to see if their wide great wings have warmed enough in the sun, then land and sit in a row again. Perhaps, like their ancestor lizards and snakes, they need more heat to get them going.

There are feathers here, white and short; a pigeon perhaps hawked from above or foxed from below. No carcass, no head or feet or beak. An exploded pillow without its case. A small breeze drifts them across the meadow. There are those hereabouts who love nature. Who ride horses after the fox, or shoot the crows from crops or pigeons for the pot. Here a death means other lives.

Pelargoniums

In the pots are pelargoniums,
flashing frilly pink and green.

We usually call them geraniums, but that isn't what they are. These I grew from cuttings last year, and in a month or two I will take cuttings from these plants and put them in the greenhouse to overwinter. They are all red. I only grow geraniums that are red; it is the best colour for them, and if I want other colours I grow other things.

The garden is looking its absolute best right now. The roses are perfect and deadheading them will go on until October, then unlike many other gardeners I'll leave them until early spring. The cosmos are nodding away, two and three feet tall with pink heads, and they too will soon need deadheading; they'll flower too until October or November. In the beds are the hydrangeas, pink and blue and white, the calendula and the lilies, the hebes, campanula and phlox, and on the walls the white climbing hydrangea; over the fence the *Solanum crispum*. In the lawn the *Magnolia grandiflora* is packed with massive velvety creamy flowers the size of dinner plates. In the cottage beds, nigella and the tall and slim *Verbena bonariensis* that I planted last year. And everywhere by the hedges there's borage and bees. My life is measured in flowers.

All there is for me to do is cut the grass and deadhead the roses and cosmos, and then with my pocket knife cut off some of the straggling iris rhizomes. The ones I potted up early last month have rooted and I use them to fill some gaps; back in their place they look neat and tidy. Nearby the dahlias are massive and crowded, just the way I like them, so that they support each other and hide the twigs that I put in between them to hold them up; there are blackfly on the stems. I do some deadheading and, in this heat, that is enough work for me. I sweat and slowly water in the newly planted iris rhizomes. It is twenty-five degrees in the garden. I do not like the heat – I am a northerner, I like the cold. Miss Cashmere seems not to like the heat, either, as she does not appear in the garden for several days.

Mowing the lawn, I am in my perfect element. A ladybird lands on my arm again; there's a tiny spider there, too. A bumble bee looks for flowering clover, and I have to stop the big mower that would suck it up and mangle it beyond repair. Every time I let go of the dead-man's handle, the mower shuts down. I usher the bee away and go back to the mower, take the handle and starter rope and start the hot machine back up again; it pops and struggles, the fuel evaporating before it gets to the carburettor, because of the sun beating down on the hot petrol tank. It takes a few pulls to get the engine going and for a moment, as I sweat and struggle to restart the mower, I decide it was worth it, just to be the one who stopped my work so that the bee could go on its way. Over in the meadow that we fill with wildflowers there are hedgehogs, field and wood mice, a pheasant and her chicks. In the swampy edges of the stream below the ever-flowing spring the tadpoles are growing legs. Dragonflies cruise among damselflies and bees and hoverflies, swifts and sparrows and tits and, above them, hawks. To sustain these levels of life the ground must be as teeming with insect life as the air is. The land below and the air above the meadow are a thick soup of life.

Gardens like these cannot last. They are messages from a bygone age and need skilled labour that few are willing to do, and fewer are willing to pay for. Even this wildflower meadow is an artifice, mowed in August with a scythe so that the flowers can dominate the grasses. If left, it would revert to grassland for a while, then the beginnings of woodland, then the deer would return and take over, cropping the grass and the flowers and saplings, and turn it into close-cropped lawn again; perhaps, if the extinct wolves returned to manage the deer, this land would in time revert to the forest it once was. Wolves create forest.

Over the pond too there is a cloud of small insect life. The water level has fallen over the years. Most liquids contract in size as they freeze, but water expands; if it didn't, it would sink and the Earth would be a dead ball of ice floating in cold space. Over winter after winter, the ice in the pond has expanded and pushed against the stone walls until they cracked, and water leaked into the thin cracks, then froze again and made the cracks grow bigger, and now the walls leak slowly. In heavy rainfall the water level rises for a while, but over several days it falls back down to the level of the earth, and where I used to sit on the wall and dip my hand in the water, now I have to reach down to where the carp live. Nevertheless the pond is filled with living organisms and I need to take a net and remove the thick layer of weed that constantly threatens to take over, leaving it on the top of the wall to dry in the sun and allow the insects to crawl back and drop into the water twelve inches below. I suppose I should repair it, so I take out my notebook and among the notes I've written in this garden, after the words 'allow the insects to crawl back and drop into the water twelve inches below', I write: 'Buy stuff to repair the pond.'

The toys are still on the wall, dry now, the mud cracked and easy to break into chunks and brush off. I pick off a few chunks of mud and leave the toys there, cleaner. They are a gift, a test to see if she notices them and mentions them, to see if she has been down this way.

Over by the hedges, tall nettles are attracting butterflies and hiding, as they do, all manner of old broken stuff: bits of fallen fence posts, and piles of stones and bricks that might come in useful some day. In the hedgerow trees above, a thrush warbles and I wander down the mown path by the hedge and past the meadow, feeling old and young at once, towards the ever-whispering aspens, whose tiny leaves are restless and chatter even in the slightest breeze, to feel a little cooler in their shade. In the evening hedge a fox slides by, low like smoke, looking for ground birds and mice to eat. We both stop and watch each other for a while; he is as entranced by me as I am by him, then he unfolds and mistily slides away into the shadows.

Flying Ants Day

In the village the swifts are circling around the bell tower and feeding fast over the black slate roofs that already radiate bright sunlight and heat. Jackdaws squabble and raise their young in the chimneypots, and Peggy and I lean on the window ledge and watch. There is not much else to do; we are in summer's little winter, when heat stops everything from growing – the streets, the gardens, the earth all around is dry. Rain, when it comes, will start it all growing again.

Seagulls are stomping on the pavement, acting strangely, and as I look closely I see millions of flying ants emerging from between the paving stones and the trodden, irrepressible dandelions that grow there: males and young queens looking to mate on the wing and form new colonies. The seagulls get drunk and stupefied on formic acid as they feed, standing around, docile and stoned. The ants crawl from their sandy paving-stone gaps and take to the sky in the form of a swirling black cloud, so we close the windows tight and keep them shut until they pass. By the evening the black cloud has passed, the flight is over; dozens of big queen ants wander around, their wings dropped off after mating, looking for a home and a new colony to start. Children stand on them to squash them while their mothers call them in for dinner, and abandoned, useless wings litter the pavement like black funeral confetti – tiny fallen leaves, teardrop petals among the squashed bodies.

Ants aerate the soil, turning over as much soil as earthworms do, tilling and distributing nutrients, scavenging and recycling fallen organic matter, distributing seeds. Some ants farm caterpillars as they do greenfly; some caterpillars, too, produce honeydew and the ants take them to their nests, where they complete their development. Ants are both damaging and useful, spreading and protecting greenfly, but also protecting plants, spraying formic acid at pests that are no use to them.

The greater usefulness of the little creatures reminds me of someone I met at a book event. I was talking to a group of people about moles, and she asked me what moles were for. She believed, she said, that each creature on the Earth had a purpose. I asked her what a cat was for, and she said it was to catch mice; and cows, she said, were for milk and meat. So I said that moles were not 'for' anything, they just 'were', but she wasn't satisfied; they must have a purpose, she said, but I can't think what. I proposed that we both believed completely different things, and wasn't that a wonderful thing? But she thought I was just plain wrong and didn't know enough about moles. I wonder, as I write, how she would have answered if I had asked what she, or I, were for.

Hoverflies hang stationary in the air and look into your eyes. I once met somebody who had decided they were alien drones, sent from another planet to watch and record us. They beamed the images of my face and hers back to their home planet; she wasn't sure what for, but was concerned about it anyway. We sat on a park bench and chatted. She spent her days looking for money, shoplifting and working as a prostitute to fund a heroin addiction. I've known a few heroin addicts, they work harder than many other people I have known – harder than ants. All day and most of the night is spent either trying to get money or trying to get heroin, or sweating and being afraid, edgy, calling and waiting for dealers, taking heroin, then starting again the next day looking for money. It keeps them busy and when they stop they don't know what to do with themselves.

Swifts Leave

On this green stage a war rages, watched by those solitary alien hoverflies, sentinels keeping post while blackbirds eat caterpillars, who eat cabbages that eat the meat of dead blackbirds and leaves. Roots wind around the old tiny thigh bones of shrews, while I look into my phone, which perhaps looks into me while I tweet a picture of a dahlia to the world.

I watch the swifts leave the bell tower and tall stone buildings where they nest and swing around each other chasing insects, weaving in and out, then leave for Africa. I saw them go at the same time last year. The young leave the nest and then, after they are gone, wave after wave of swifts come from the north, passing through for several days. They come to breed, then two months after they arrive they fly with their young back to Africa; the young have never flown before. Silently I watched one of them and tried to hold the pattern that he made as a wandering line across the sky. I know when they are leaving because they fly away, instead of around. A faint pencil-drawn line that faded and strengthened, as my attention did. A distant radio signal from a ship at sea on a stormy night reminds me of the day I watched this happen, a year ago, when I decided to start writing this book.

The young will stay aloft for three years until they come back to breed. They live for twenty years, and in that time will fly the equivalent distance of going to the moon and back seven times, mating, eating and sleeping on the wing. They do not rest. They never rest. Their wings are so long that they cannot land on the earth; if they did, they would not be able to take off again. Their feet adapted for clinging to vertical walls, they only touch the earth in death. If you see one on the ground, you must pick it up and throw it into the air. I had thought swifts were related to the swallows I used to see in my childhood, but they are not; they are more closely related to hummingbirds. If you find me on the ground, stuck to the earth, please throw me up again.

Excess apples are falling from the trees, hollowed out by wasps and young blackbirds. A smell of cider hangs around, as they start to ferment with their own natural yeasts that grow on the skin. Miss Cashmere's tortoise-shell cat jumps on the wall that runs round the pond, stepping prettily; she comes to the line of rusty old toys and pushes each one, individually, into the pond with a little 'plop', where they will decay and be forgotten again. Cats are Buddhas, and this one is clearing the past, getting rid of the dust of pointless attention.

Miss Cashmere comes down, and I see her stop and lean against the wall as if she has had enough, then bravely she gathers herself, un-leans and moves on away to the summerhouse, like a blue tit to her nesting box, carrying cigarettes and newspaper – nesting materials that will bring her comfort and make her feel she still exists in the world. She is now more bird than human, from a distance at least; her presence is a singing, chirping colour that doesn't need to be, but is. She lasts a moment that lasts for ever. She is hard-edged, like a Western painting, filled in neatly between the lines. Even though she is frail, she is sharp and neat, like young pines against a clear sky swayed by light breezes, while I am smudged and foggy, like Japanese ink painted on absorbent paper. I spread a little at the edges and soak in, mist and rain and blur.

Pine Cones

I find a group of pine cones, green and unripe, not yet open, fallen to the gravel path. I take a handful to a garden sculpture by the house – a piece of white limestone carved into the form of a giant pine cone – put them on the plinth and think how skilful man is to make a permanent copy of something designed by nature to fade and cycle round again. This poor fake will never fade. It lacks the detail and subtlety of the real thing. I wonder what it's for. The moss is starting to grow from some of the deeper cuts on the shady side, where spores landed and found moisture. Perhaps I misjudged it, perhaps it is alive after all. A bone growing some flesh to cover its nakedness with.

Carp

One of the enormous carp from the pond lies dead on the lawn, still fresh. Half-eaten, a strange hybrid creature of mouth and eye and gold-green scales like tarnished coins, a handful of mixed change, that abruptly alter along its length into a bloody mess of thin white pipes and strings attached to purple bean-like organs and layered fibres of pink muscle, pink gnarled bones and tiny white needle-bones, then more red bloody flesh and overlapping brassy scales, green-hued tail – a lovely mess. It looks like it should be lying just as it is on an oval dish with slices of lemon, at the centre of a table surrounded by people looking at it, licking their lips and holding knives and forks. A bird or fox, possibly a buzzard that leaned over, feathers, beak and claws, to dip in, dip in and pull the thrashing life out of its world and into the heavens, has eaten the middle section, leaving the head and the tail. It has the outline shape of a cartoon fish eaten by stray cats in an alley, but is two feet long and really there in pasty, wet reality. And there it stayed, because I left it to see what – if anything – came to claim it later. Later came along and it was still there.

There are a number of large carp in the pond. When they were introduced many years ago, they were already old and had reached the limit of their growth, but they were caught and carried to this bigger pond and almost doubled in size – the biggest now being well over two feet long, maybe three. Perhaps they are magnified by the water and appear bigger than they really are; some things should only be glimpsed. The harder you look, the more things change. Things grow to fit the space available: we expand, and we contract, depending on our circumstances. I was a much smaller person when I arrived in this garden, my outlook narrower, my body slighter, and I expanded into it over time. When I plant the dahlia, I become the dahlia; when I focus on pruning the tree, I become the tree. When I'm not focused, I become the spaces in between. Sometimes I feel the garden is indistinguishable from myself.

On a still day, when the water is clear, the fish can sometimes be seen cruising about near the surface, and if Miss Cashmere is not at home I'll rest and eat my lunch on the stone wall that surrounds the pond and watch them in their own inaccessible world, appearing and then fading as they emerge, then slide back into the murky depths. Throwing crumbs in the reflective water that looks like the sky, I try to look beneath the ripples and reflected clouds to watch them congregate and cruise, this way and that, near where I sit. They taste the crumbs with their big mouths and, pushing their hard lips out on weird hinges, spit them out again. Sometimes one will suck my finger if I trespass it into their world, like the hand of a god coming out of their chaotic heaven. This dead one had been dragged into my thin, unresisting world and died here on the grass. I wonder if this individual ever sucked my finger.

I live mostly alone in this garden. Like a cow treading on the thin, flat layer between the chaos of the thick sky, where the birds and insects defy gravity and fall in all directions, and the deep earth, where the moles and worms and mycelium, much slower in their more resistant world, push soil and burrow in any dimension like slow fish, I decide that just at the moment I would rather be a bird in the vast emptiness than this man-cow on my paper-thin slice of world, and definitely not a fish cramped into that murky sludge full of parasitic bugs. Yet man-cow I am, horned and hooved and roving across my flat labyrinth. I have been here so long I have probably touched and impacted on the life of every single growing thing, every blade of grass, every tree, every perennial, every annual.

Green Flames

Suddenly, and without warning, the winds have started to blow across the Rookwood where I live. The end of a storm that devastated communities many hundreds of miles away. Yesterday I noticed the leaves on the sycamore trees were turning orange. The oaks are losing their vibrant green, and I can see the beginnings of pink on some of the cherry-tree leaves outside my window. The grass needs cutting, but I could not be bothered to do it. I feel tired. It will soon be time to enjoy the depth of long nights.

I am burning dried brambles and branches and cuttings, which have collected in a huge mountain behind the stables. The last horse that lived here died after eating buttercups or ragwort or some such, or tripped in a mole-hill and had to be shot perhaps. I am a heathen, ignorant; I go blank when people talk of horses. The only horse I ever got on with was a donkey on Blackpool beach that my friend John Cooper owned, or his dad did. I have never owned an animal. I am not sure what they are for. My father often had a dog. I have a cat, but ownership has never been part of the relationship. I see my cat as a teacher, and she pretty much looks after herself and can do fine without me.

The bonfire crackles and smokes in the wind that blows into my eyes and stings, and I am enjoying the unpredictable swing between heat and cold, smoke and fresh air. Fire bathing. A green flame tells me there is copper in the fire, and I wonder if some plants contain enough copper to make a green flame or is there just a piece of copper wire wrapped around a branch? The books are gathering nets of grey cobweb and I consider adding a few to the fire, but it is smoky enough as it is. The green flame reminds me of when I was an art student; unable to afford materials or tools, I used junk that I found on waste tips and in second-hand shops and abandoned factories that were scattered around the North. I burned the plastic off copper wire on the dump. I had listened to the pinging of Morse code on the tiny transistor radio that I had been given one Christmas as a child, and would listen to ships at sea through the night to keep myself company. Instead of sculpture, I made radio receivers from books and branches and granite doorsteps where people had passed, and electric batteries from foil and paper, lead and zinc and copper scrap, lemon juice, vinegar, sour beer. I wound miles of copper wire from the rubbish tip around stones and the branches of living trees, connected the earth to the air with metal spikes and, with salvaged earphones and loudspeakers, I tried to listen to the voices of the dead, but all I heard was static hiss.

One per cent of all static hiss on our FM radios has travelled through space and time from the world's first day, the very beginning of each mote, planet, snail, eyelash and word that exists today – the sound of interference from the Big Bang. I didn't know this at the time, but I did realise that I was not making sculpture any more. I was looking for voices from beyond, so I went back to writing poetry in the gaps between frying chickens for drunks in the chicken shop where I worked at night to pay my way, and I learned to carve words in stone. It was always words. In my loneliest moments, books and the people in them had been my friends, my family, teachers and lovers. I just wrote; I wrote to feel and record and amplify the breathing rhythm of the world, to make it pound closer to me. Poetry is the only way to see what's on the other side of the veil. The vagueness of this language of birds draws the fragile human soul out from its hiding place to brush against, with its slender form, the vapour of the real world, and there is no other path between the earthly things and spirit.

August

Cofiwch Dryweryn
(Coffee-ookh Dre-weh-rin)

The hose only stretches as far as the dahlias, so I carry water from the barrel to the young hydrangeas that are scattered in beds all around. My shirt is sticking to the sweat on my back, and I feel the skin on my head tighten as I start to burn. There's humming coming from the lavender, and toing and froing on the ant path by the gate. I carry two ten-litre plastic watering cans. One litre of water weighs 1 kilogram. Up and down the garden I carry the equivalent of a two-year-old child, although by the time I get to the plants some has slopped over onto the dry earth. Buzzing fills the air. The long days start and end with sun. The hydrangeas are wilting; they are the first shrubs to start looking stressed in times of drought. I pour gallon after gallon onto the base of the plants, and within hours they stiffen their drooping stems, lift their large floppy, wilting leaves to the sun. But soon they are falling again.

It has been hot for weeks. The grass has turned gold and stopped growing. Leaves on the ash trees are curling and dropping to the ground as the tree tries to prevent water loss. The lettuce, thinking they are dying, have bolted and instead of the tight shiny green globes, they are now tall plants with stems and leaves and flower buds at the top. The lettuce wants to flower and set seed before it dies. There is just me and the plants and the birds and the sky all over us. The butts are nearly empty and the water in the taps tastes murky, as if the reservoirs fifty miles away up in the mountains are getting to the bottom and I am drinking filtered mud, the stuff of life. Fragments of the past are found there; some futures end there. The village of Capel Celyn might be visible in the mud, now that the water level has fallen. The graveyard and the chapel walls. The ancient buried dead were dug up from there and moved away from their square mile before the Tryweryn valley was flooded, and a Welsh-speaking village of people were removed from their homes in 1965, the community destroyed, their homes filled with water. There are photographs of crying children being forced out, carrying their few possessions. Country people are very attached to their land and their language. Outside a village called Llanrhystud there is an abandoned cottage wall that faces the road; it is painted red, with the words *Cofiwch Dryweryn* painted on it in white. 'Remember Tryweryn.' It is one of the few monuments to the Welsh there are. Every few years it gets vandalised. The community repairs and repaints it.

I am not a country person – I do not have a country. The languages I can hear from where I sit writing are just a few of the languages of the world; the people speak three that I can make out. I can hear the birds speak another four, the tower bell speaks another. I understand some of these calls; others are music played on different instruments from my own, they sing the songs of the morning. Language comes from place; it is the most powerful thing that displaced people have that is their own. Until the 1940s, schools here in Wales had a practice of punishing children who spoke the language of their home, their parents and family. A child heard speaking Welsh to his friends or brothers or sisters in the playground or classroom would have to wear, on a string around his or her neck, a wooden plaque with the letters 'WN' carved in it. He would be encouraged to listen out for another child speaking Welsh, who would then have to wear the 'Welsh Not' until another child was heard. Whoever was wearing the plaque at the end of the day would be flogged by the teacher. There are many tales of brothers asking their excitable playful sisters, in Welsh, to be handed the Welsh Not at the end of the day, to save their sister from a caning. It is still very much a living memory here; the old people remember wearing it and find it difficult to trust the English. I'm a Gipsy, a heathen, an anarchist without a square mile of my own to protect, so nobody trusts me anyway. The language is growing back again, strong and prouder than ever, after being pruned to its roots. And although I don't understand a lot of it, I love to hear it pass my door – a living thing. We don't need to understand in order to love.

Umbellifers

Over the meadow there is a mist of insect life, the air is fogged with their millions. The constant hum and buzz of summer. The muddy patch beside the thin stream is dried and cracked. A billion butterflies forage among the yellow, white and purple flower heads of thistle and bellflower and horehound and borage. Some of the flower seeds in the meadow are dry and leaving home. In a month or so it will be time to cut it and lay it in windrows for the seed to fall.

Miss Cashmere comes out in a summer dress; she is neat and coloured with something orange, from the Fifties perhaps, that has come round to fit her once again. I'm mowing a path around the edge of the meadow. I have mown a path there for years. I haven't seen her use it for a long time, but I still mow it; we pretend that everything is normal, that the path will be used again, that everything in the garden persists, perhaps for ever. I turn off the mower as she is coming towards me. It is hot and she is slow and it takes an age, but I do not go to her, I don't know why; it would feel as if I wanted to see her about something, instead of her wanting to see me about something. I let her walk – perhaps she feels like walking.

Her ginger lap cat hasn't been seen for days. 'I know he is only a cat, but I miss him,' she says.

'I'll keep an eye out for him,' I say. We talk about the heat and she thinks it is wonderful, and I tell her that 'I prefer the autumn'; that 'I come from the North and enjoy the darker, colder months.'

And she says, 'Oh, how funny.'

I tell her that 'The pond has a lot of weed, and perhaps it might be an idea to turn on the fountain to aerate the water', and she says that she will.

She just wants to talk, it seems. 'There are lots of plants here that I haven't noticed before,' she says. 'Do you know what they are all called?' she asks.

'Mostly,' I say and tell her about the sedges that grow where it's damp, and the yellow rattle that feeds on the roots of grasses, and the campions and cow parsley. And she asks me if these other umbellifers, with their wide flat flower heads, are baby cow parsley. I bluster and say, 'They're from the same family' and hope the name will appear, but it doesn't and I feel stupid. 'I can't remember what it's called,' I finally have to admit.

To me, it doesn't matter any more. I know what they need, what they look like, how they behave, where they like to grow and where they don't like to grow; it is all obvious from looking at them. But I cannot drag their name from the rusted-up drawers in the basement of my head. I could draw them with my eyes closed, because this is my daily life. I could tell her that the hoverflies and butterflies love the tiny white flowers. I could tell her that they come from the same root every year and like the wet, boggy parts of the meadow. I could tell her that they smell a bit like cat-piss and a bit like elderflower; that they come from a small green fist in the ground that opens into a flat creamy crown of flowers bunched together; and that in the winter they will remain standing and brown, with a head of flat seeds that the sparrows love to eat. I could tell her that it is candied sometimes and used to decorate cakes. But the word will not come.

I could simply pull out my phone and search for 'Umbellifera', for they are obviously of that family, which includes edible plants like carrot and parsley and sweet cicely and wild parsnip; and also includes hemlock, which, after eating only half a dozen leaves or a few seeds or piece of root, will make you feel cold at the feet; a paralysis then slowly creeps up your body until it reaches your lungs, which collapse as you die. But it is none of these and I am silent and aware that I might look foolish. I finally tell her that it is wild carrot.

She looks at me and she says, 'Wild carrot' in a quizzical voice, as if either she knows that this is not wild carrot and I am lying, or she does not believe there is such a thing as wild carrot. It is four times bigger than wild carrot. Nevertheless, she goes off to the summerhouse and I fear that she will look up 'wild carrot' when she gets back and see that it is similar, but not the same, and I feel guilty for misleading her so that I wouldn't look stupid. I don't mind if I look stupid when I am not there, but I would rather not look stupid to her face. I feel ashamed for the lie, which appeared already formed and ready. Then as she goes slowly away, too late to do anything about it, I remember that it is of course angelica.

I lean over, pull the cord and restart the lawnmower, cutting past a patch of red corn poppies. The field of poppies and mixed wildflowers is astonishing; they suck the oxygen from my lungs and force me to let go of the mower, let it sleep while I stand and look. I am catatonic, paralysed, open-eyed and absorbing it. They are all the peoples of the world ageing together. The brown and pink and red and yellow people, the tall and short, the fat and thin, the gay and straight, the Catholics and Muslims, the fancy and the simple and all the spectrums in between – all using similar nutrients, all breathing the same air, drinking the same water. I feel okay about forgetting their names, for labels are such temporary things and rarely capture the enormity of the thing they try to catalogue.

This moment is the best of my day and I drink deep. The ginger cat is chasing butterflies, and an old stray tomcat, filthy, injured and abused, slinks along with pride under the hedge and, seeing me, that beat-up-looking Buddha goes back again and fades into the long grass.

Fountain

The heat drains me, and I have been looking to the sky for relief. In summer the garden is not very demanding. The whole of the rest of the year has been about preparing for these two or three months, and the glowing autumn that follows them. Now I can slow down, take my time, seek shelter. At least today feels cooler and there is some thin cloud over to the east. There may be a shower later this week. The lawn is yellow and hydrangea leaves are wilting. I take out the hose and soak the dahlia bed. There are blackfly on the stems, feeding and being born, and in a mean mood I direct the jet of water to blow them off and make them work for their meal. The sun is behind me and it makes a rainbow in the falling spray.

Miss Cashmere sits in the shade of her house with her cigarettes and newspaper, drinking from a teacup, then rests it in a saucer next to her ashtray, a lumpen brown thing, possibly made by one of her children. The summer-house, a glass-fronted wooden box, gets uncomfortably hot and she hasn't been using it for a little while.

I move slowly towards the pond and catch her attention, waving at her and pointing at the fountain, and she remembers that I suggested turning it on. She flaps a hand at me, as if to tell me that I am a nuisance, gets up, hobbles to the wall and turns a switch. A moment of gurgling from the fountain, a trickle, an ooze of green slime that slowly clears to a single jet, growing more forcefully into a single umbrella of clear, bright water sparkling six feet into the air, which falls and splashes on the dry bronze decoration, washes off the dust and dirt and reduces the air temperature noticeably for a few feet all around. I give her the thumbs up and she goes back to her newspaper. She has told me that the fountain is expensive to run, so is only usually turned on for special occasions or when guests are coming. It hasn't been on since this time last year, when the pond weeded over. It is becoming a routine in the annual season of drought that this tremendous geyser is sent up, to glitter in the sun and taunt the parched earth and its sweating gardener with its unattainable coolness.

I stand for a while with the sun on my back and watch the falling drops that patter on the surface, as they re-join the pond and are then sucked down through the pump and thrown back up again, a few reaching the earth, leaving the pond perhaps for decades until they return as rain. If there were a breeze, the spray would blow and I would drift into it, but there is no breeze, the air is still and hot. I remember a little child's blow-up paddling pool, the smell of hot vinyl, a blue base and two inflatable tubes making the wall, a yellow one and a blue one; the garden hose spraying, making a rainbow over . . . who? Was it me and my brother? Was it my own children? Was that my mother, tall and smiling, leaning over? My father spraying us with the hosepipe? I push the image away. It is a tiny loop of film that goes round, and the universe to be seen in that particular grain of sand is just leftover stuff from the past. Dust that I brush away, as the cat brushed away the rotten toys. I have much more enjoyable ways of wasting my time.

Cats and Dogs

Miss Cashmere has two cats, I have one. I like cats. I think of them as very calm creatures that like to have fun, but are sensual and know how to relax; they don't waste energy, and time spent with them is never wasted. Developing a calm mind is like building a relationship with a cat. If you try to make it come to you, it will run further away; if you chase after it, it will hide; but if you sit quietly, keep an eye on it and appear ready, it will come to you, and it may stay or it may go, but a relationship will have been created and a stronger one is more likely, the more you sit and practise.

When I was a boy I had seen a picture of St Francis of Assisi in a school Bible and I really wanted to be like him, to have the wild animals feel comfortable with me, the birds sit on my shoulders, the deer nuzzle me and share my space. I fantasised about it and I practised keeping still and calm and waited for birds to land on me.

One day, coming home from school when I was about twelve, there was a group of children running around terrified, screaming and jumping up onto a wall to escape from a nasty dog. There are always tough kids in any school: kids who have rough dogs, bullies who fear gentleness, lost souls who laugh at the fear they create in others. These two brothers had brought their vicious white dog to the school and let it off the lead to attack the children; it was chasing them, snarling, baring its teeth and snapping at the kids, who were trying to get to safety. I felt deeply that if I remained calm and peaceful, it would not attack me. I believed that the running away and screaming were encouraging it to chase them, so I stood calm and smiled at the dog. I had grown up with fear and knew how to handle it. I refused to run – I was not going to behave like a prey creature. I didn't make any sound whatsoever, just held my ground and avoided eye contact.

All the training that I had given myself up to this point, in being calm and still, not giving in to fear, was about to come to fruition. The dog bit my leg into red dangling rags and later on, lying in my saintly hospital bed, I vainly tried to figure out where I had gone wrong.

Distant Sounds

The month is in full flow. August seems to fly by, when other months drag on. Summer crawls in slowly, then at the end skips out too fast and, as my head turns to watch her dance away, the cold creeps up behind and makes me lonely. I mow the path around the meadow and some of the wildflower seeds are leaving their pods in the meadow already, travelling as far as they ever will and landing in cracks where they'll root and spend their lives a human stride away from where their parent was born and died. And next summer they may be fertilised by the very same bee that snuck into their parent. It is very nearly time to cut down the meadow. I need a few days of sunshine to ripen and dry the stalks a little, for it is impossible to cut wet stalks properly, and very dry stalks cut badly. In shorts and straw hat, I inspect the meadow to see if I can cut it yet. The low morning sun lights up a million glittering, dew-sagging spider webs and a gaseous cloud of tiny insects covers the acres; the golden-brown meadow is alive with minuscule creatures and I cannot bring myself even to think about cutting it down, so I sit on a stump and watch. I am caring less and less about being caught not working.

It is hot and the garden is painful with colour, the brightness of a million smiths hammering resonant metals, clanging copper and brass into petals that turn to catch the sun and burn the eyes. The back of my head hurts with squinting, so I head to the shady dampness of overhanging trees and ferns and leaf-filled pools that bubble with decomposition. The silence and dark are precious and profound with beginnings. Slowly I become attuned to the darkness, and the browns open up into a million shades, from yellow through orange to red. Distant sounds come in and over the garden fence; the summer hawks circle the blue above a field where farmers are sweating in tractors, teddering crops that have grown dry and golden in the sun. Spinning metal fingers worn to a high polish fling the stalks about, as men would have done not long ago with pitchforks, so that the warm air can get in to dry the stalks. Active, creative men who smell of the earth, earning their keep with tools. Black cows lie in the shadow of an old chestnut tree and, as the sun moves the shadow, the cows at the sunny edge get up and wander to the far side of the herd and settle, reshuffling themselves and drawing the shape of the tree moving slowly across the ground.

It is too warm to work and there's nothing to do anyway. Summer is middle age: very little happens, the kids leave, the mortgage gets paid, young people overtake us, and we remain and wonder how to spend our time. All we can do is sit and watch it go by. We slowly mature; we are a cheese made and resting in a damp cellar, waiting for the flavour to develop. We are whisky in the barrel, the brash, cheap sugary summer waiting to become rich brown autumn. I can sit back and watch the world perform its summer song.

Pond Scum

In the border by the drying lupins and foxgloves I'm kneeling pulling weeds, bare-handed in the soil with dirty fingernails and gritty hands, and find a beast. In my palm, as long and as fat as my third finger (and my hands are big, a gardener's hands), an elephant hawk-moth caterpillar, brown and grey with massive eye-spots, rears itself defensively and almost hisses. He was looking in the leaf litter for a place to settle over the winter while he completes his transformation into one of the biggest moths we have here in the UK. I gently set him down and bury him with litter as he ambles off. Beside me, a robin has spotted him and I shoo him away. I notice my inability to resist interfering in nature, and wonder about how I came to do a job that mainly consists of doing exactly that. In reality, I don't think I am suited to any job at all. I have had so many different ones and always eventually find them ill-fitting: I do not follow the rules.

When people ask me what I do, I say that I am 'just a gardener'. I feel that I am part of the landscape, but Peggy gets annoyed with me for saying 'just a gardener'. She says there is no 'just' about it; she says it is an honourable profession – an honest and worthwhile one that pays the bills. I don't want to be anything else, but still, my bones ache. I am a solitary person, and my deepest relationships are not with humans, but with wind and rain.

This is an everyday domestic house and I am an ordinary gardener. It has no history that I know of, the garden does not win prizes or open to the public; it is commonplace and magnificent. On the stable where I keep my tools, a rotting thick wooden door with traces of pale-blue, green and oxide-red paint etched in the deeply grained splitting wood hangs on a rusted orange hinge, a type made by a blacksmith long ago, and beside it a downpipe green with algae that has taken decades to grow its abstract, shaded patterns. On the stone walls, yellow and orange patches of ancient lichen.

On the pond and in the ditches there is a thin layer of green scum. This common green scum is a community of astonishing creatures – billions of them. Scientists can't agree on whether it is a plant or an animal, and it behaves like both. This single-celled organism photosynthesises when the sun shines on it, converting sunlight to sugars. When the sun moves and the creature is in the shade, it whisks its tail and spins until a dark patch on its body casts a shadow over a light-sensitive spot below, and then it rests and photosynthesises. With its pirate eye-patch, it follows the sun until dusk. At dusk it is aware that the day is ending and it stops looking for the sun; it rests and absorbs carbohydrates by osmosis, like a coral or a sponge. This plant / creature is euglena (*Euglena viridis*).

Euglena comes in many varieties, and it is pretty much everywhere on the surface of this planet. If I put my hand in a ditch and scoop up a handful, I would be holding millions of them. They would carry on living their lives just the same, unknowing that I held them. Overhead an aeroplane pumps out fumes to kill it all, as people go to Kenya to see larger distant beasts.

Laurels

There are no eagles here, no rare nightingales or endangered orchids; everything is commonplace, yet all the commonplace things are extraordinary, in that they exist at all. The *Euglena viridis*, with its sexy pirate eye-patch, and the peppered moth and the cherry tree and Miss Cashmere and her cats – all are extraordinary and magnificent through the simple act of being here and living, when for everything that lives, ten billion things did not make it this far: things that did not sprout, seeds that landed on tarmac or bed sheets, or fell short of the egg. The six-week-old mole that left the nest to start its own independent life, but was immediately eaten by a crow; the baby born with Cyclopism that lasted just a few days. Each of us the miraculous outcome of a billion chance events.

The jackdaw young have fledged, but they stand around and flap a bit, still moving up to their parents to be fed, who hop off, squawking. There is another wave of swifts, whistling high and rapid – are they the same swifts or a new batch passing through? They are not visiting the eaves; the young have gone. They fly low to catch dawn's damp insects, rising for warmth in the shadow between the houses. The usual row of grey pigeons, making bubbling pigeon-sounds on the same roof ridge they sit on every day.

It's August and time to prune the laurel hedge, as I have done every year in August since I came here. I trim it with secateurs, clipping the laurel branch by branch; if I used the petrol hedge-cutter the leaves would be smashed and ugly. I like to prune things so they look as if they had grown that way, cutting hard behind a leaf that hides the bare end. As I cut, I can taste the scent of bitter almonds. The leaves give off cyanide, which amateur entomologists shred into their killing jars, for killing moths and bugs and butterflies; the professionals have access to far more lethal poisons.

A Break

After four solid weeks of sun and heat I am exhausted, and pleased as I open the blinds to see a misty rainfall softening the buildings and the sky, and the silhouettes of newly fledged jackdaws on the chimney stacks, who stretch out their feathers into the first rain they have known in their lives. I can smell the wet earth and can't resist going out before the day has properly started. I take a short stroll in the rain down to the village café and drink a milky coffee, looking out to watch the villagers pass with umbrellas, old people going to the post office (there are always old people in the post office, as if they run on rails from their sheltered housing and back).

Graham the postmaster is grumpy, and the villagers all know him and his moods. We live in fear of having to go and face him, he will invariably tell us what we have done wrong: address too large, so there's no room for the stamp; or too small or badly written, so that the delivery person will struggle. He asks the contents of every parcel, because 'The law says we can't send certain things', or he'll comment on the scruffy packaging of a parcel, saying that it will be in an even worse mess by the time it arrives, 'Won't it, lovey?' He announces all this loudly, so that the other customers smirk in the queue at the one being told off, their smirks fading as they get closer to the counter and they look nervously at their own package. I was once very audibly chastised for my handwriting, and another time for sticking a stamp on upside down; he said it was against the law, but 'Well, it's stuck down now, isn't it, lovey?' and he threw my postcard disdainfully into the sack behind him.

There are children in red cardigans and coats being taken to school by hurried parents who need to get off to work. I wander around the village to see what's going on. Buddleia, blue and browning, perked-up hydrangeas and soaking roses bowed down by the weight of water. Birds singing from the aspen trees. Dill seed ripening. Clear water pours off slates and down and out into the street through galvanised downpipes, trickles, then flows and then trickles again as it washes; floods, then thins.

How can I write the stories of these days? They go as fast as they come, and I am after them with my net to bring them in, and hope they will flutter or crawl or somehow land on these pages. Slam the book closed now, so they don't escape! They are small and imperfect, like us all.

Gathering Seeds

In August there is a tiny growing space in me that is painfully aware that autumn is coming; it brings the sadness of endings and the joy of beginnings. They both come from the frailty of living things. In the middle of the brightness of summer I become aware of the shade of death. We know that there is no beauty without impermanence, there is no light without shadow, yet there is a sadness that comes from not knowing how to keep the things and people we love from dying. A sadness that is itself a lovely, delicate thing.

The seeds are drying on some of the annuals, and my pockets at this time of year are often filled with seeds: marigolds, poppies and nigella more often than not. Poppies have seed pods, which are the swollen ovary of the flower that sits on the dried stalk, and they come in a million different shapes and sizes. The Welsh-poppy seed head is just under an inch long, narrow and, as it dries, it starts to open from the top, the sides roll down in between stiff ribs and as they roll back slowly, the tiny seeds inside rattle in their little pepper-pot and the breeze blows them away. As they empty, the sides fall away, leaving an empty cage of ribs that will fall to dust by the spring. Other poppies make 'pepper-pots' too, some big and round and globular.

In the flower beds the earliest flowering calendula are seeding, claws of green seeds turning brown amongst the bright flowers. As the flowers fade, I have been snipping them off to stop them producing seed, so the plants go crazy and make two more flowers and twice as many seeds, until there are so many flowers that I could not possibly deadhead them all. I leave them to seed, and the seed dries into clusters on the end of bare, rattling stems. I take them in my hand and, with a tug, they fall into my palm. It is cool and the smell of autumn is in the air. I am wearing my tweed jacket and I put the seeds in my jacket pocket, as I always do. There will be seeds of all varieties of things trapped in the lining of this jacket that go back years. I can't possibly collect all the seed, and enough falls to the earth to keep this little bed flowering every year. I fantasise that one day, when I am too old for this labour, I will lay my tweed jacket on the ground and, as the wool rots and turns back to soil, all the seeds trapped in the seams and the lining will start to grow in the shape of the wool jacket, as it is consumed.

September

The Waste Land

The mornings are bright and cool, with a red sun that cuts through the trees and sometimes a mist that burns off by eleven o'clock to leave the day hot and clear.

Drops are running quickly down the panes and, reading for a while, wrapped in my favourite tartan blanket in the kitchen by the streaming windows with a mug of hot chocolate, I watch the broken dawn arrive and decide to stay at home and read all day. The garden will be too wet to work in. There are jobs that could be done – cut logs, stack logs, oil and sharpen some tools – but the motivation is not there. I have to work to eat, but I have to not work sometimes, too. My mind, my body, needs the space to rest.

Before the dawn is a good time to read poetry. The world is emerging. The mind is quiet as it emerges, too. Passing the shelf on the way to the kitchen to make coffee, I pull out something at random. Today it is T. S. Eliot's *The Waste Land*: 'What are the roots that clutch, what branches grow / Out of this stony rubbish? Son of man'. Reading about a dry, waterless wasteland, from time to time I pause to think and watch the raindrops splash off oily green leaves through water-streaked glass. This is not his waste land. I can look at it from the outside, from a place of safety.

Sunlight comes creeping in, as if ashamed at his weakness. Why in the summer do I think of the sunlight as female, and in the winter as male? He is leaving and cold, drifting away now and moving slightly further each day. He will, at the spring equinox, become female and advancing. Warm and nurturing. Oh, the stereotypes! The perspectives I grew up with and fight constantly to be without. Splitting the world along such simplistic gender lines is such poor thinking. Cold and advancing, warm and leaving, male and female are meaningless; live as any, live as all.

A homeless man in town yesterday crouched, bent like a coat hook, on a dirty blue nylon sleeping bag. Standing over him, another of his tribe saying, 'You need to get back with your wife, mate.'

And the reply, 'No, I don't want her back, I'd rather be out here.'

I watch the homeless and their ebb and flow. They are our weeds, our very own wildflowers that grow in the cracks where they are not wanted and from which the tidy-minded evict them. I watch them as if their movements might tell me something, as I used to watch the crows and look for omens, throwing them a crust from time to time. To most people, they are like the animals and not really taken much notice of. But I watch because I was one of them once and know their life: ragged underwear, cold food – when there was any food at all – for weeks on end, socks that end as dirty knitted cuffs around my ankles, soles worn through, so the skin of my heels and balls of my big toes were directly on the ground like a pilgrim's, dirty and invisible and hardening into leather. My movements had a purpose, and I chose that stony ground and all the branches growing on that path, and following those branches brought me here, to love. But those homeless are weeds that are multiplying; the garden they live in is neglected and becoming a wilderness.

The night temperatures have started to fall, the leaves turning browner. Some of the homeless will die again this winter – you can almost spot which ones. They have lost their fight. They curl and wait, sucking tiny warmth from cold bricks. Hat on the floor, or tin or plastic cup from the bin outside the coffee shop, and looking down all day. They have not all chosen this life. They do not walk or become pilgrims; they stay in the crack they have fallen into. They are not ready.

'Go, go, go, said the bird'

Today's small rain will perk the grass up and it will start to grow again, then need cutting. A few marigolds are still flowering in the herb bed. A handful of buds that may or may not open. A flock of sparrows. In the rambling rose the last few yellow flowers still open, advertise themselves to birds and insects. The rosehips are turning orange. This light rainfall that will persist all day is darkening the browning dill, whose seeds I haven't yet collected. I feel the damp breeze on my bare arms, the cool damp air flowing into my nostrils, the warm damp air flowing out. The hart's-tongue fern looks varnished bright and crinkled; the bunched leaves pointing up show that the undersides are heavy with tiger stripes of umber-brown spores. The browns are vibrant dark oranges and blacks, and the air smells of wet earth.

Most things are drooping under the weight of rainfall, apart from the ferns and the nasturtium leaves collecting little balls of water, silver bearings, clear and rolling to the centre of their matt, flat waxy leaves while everything else looks soggy. The breeze blows across me and a small leaf flutters by, while I hear without listening the drop-drop-drop of water falling from the roof into its puddle, and the quiet white hiss of light rain striking millions of stiff leaves, each one rolling off the leaf to hit another. Then sparrows singing, hidden in the twisting brambly rose among the hips and the soggy last few blooms. Drops ripple in the puddles, spreading reflections.

A solitary bee searching the last few flowers. The blackbird sees me and starts its chuck-chuck-chuck alarm call, and for a moment the sparrows go quiet. I do not understand their language. Seeing that it's only me moving slow and harmless, they restart their chirping and flying from one shelter to another, in the masses of tangled orange hips that grow across the shed. A seagull flies over, and two desultory wood pigeons drip silently on the roof. I have decided that today I shall do absolutely nothing. I will read, and eat something, watch the weather do whatever it decides to do, and watch the birds.

The hours pass. As the days become shorter and the darkness comes in earlier, I begin to feel the inescapable sweet melancholy. This will last a week or a month, until the bright light of autumn proper comes and shines through the changing leaves and brings a cleansing frost. The living things start to fade and sleep. Flowers become seed pods that dry and brown, and the seeds leave them behind to fall and crumble. Waiting out the cold and wet, they endure the wind that dries, the cold that freezes and the rain that swells, until inevitably the sun shines and warms them just enough for them to drive a root down for stability and moisture, and a pair of leaves up for sunshine and air, and a plant will be born here in the soil where it fell, where it will live until it too fades and browns as its own seeds go off.

As the sun falls behind the trees, the light clings to the trunks and thin, bare branches and silhouettes them against what remains of the dimming light; the shadows seem to possess a richness that is hard to describe, as of thick fabric hiding bare flesh, the patina of age, the grime of life, an ink-wash painting – over-painted, hiding details in layers of rich darkness that makes the brightness of day and the glitter of man-made stuff seem cheap and superficial. These shadows and layers of light – as the day dies and the sun and its people fall into their night-time routines of food and home and love-making, of reading books and, ultimately, sleep – seem the most sad and the most creative, as if the darkness were where all life comes from.

The hours pass and the shadows fall, and the mysteries open up in corners where the details hide. I sit under a wooden table lamp that thins the darkness and listen to the song outside and read *The Waste Land* yet again. Of the Greek seer Tiresias, who was blinded and understood the language of birds and communicated with the dead, and was turned into a woman for seven years, then had children, before becoming a man again. I read it four times and each time a little more of Eliot's genius is revealed from the thick shadows, the mirror becomes clearer, and I find a story about a society that has lost its values, that doesn't care for life that fades, but values money and glitter and fame, over love and our five sweet senses and charming modesty; and I wonder about the corpses we plant and what kind of flowers they may grow. Our senses are truly the only thing that we have.

I cook and eat small, light food that is sweet and sour and dark, and ponder communication with the dead and the language of birds. Odin and his two ravens, who told him of the activities of men. The languages of birds are understood by gods and seers and poets only. 'Go, go, go, said the bird: human kind / Cannot bear very much reality,' says T. S. Eliot.

The Many-Forking Path

Once again I am mowing the vast lawn and making the looping mown labyrinth that folds back on itself, unbroken when it reaches a wall or a fence, and ends up just the throw of a large stone away from where it began. This labyrinth makes me think of the infinitely forking path that brought me to this place.

I am, as I write this, sixty-three years old and have lived a life of many colours. I slept as a boy alone, shivering outside like a rat and a freezing cell's width from my own death, homeless, unloved, dirty, starving. I have stolen food in order to avoid collapse and have slept under hedges with the birds. Before sleeping with the birds I'd been sliced thin; over the years thinner and more thin, until only the slicer with his knives could see me with his sharper eyes; then flung out eventually as worthless, and caught by the wind and hung up on a hedge to sleep with the creatures, who saw right through me. I wandered. I didn't wander with Gipsies like Virginia Woolf's Orlando, or be abandoned by them, like John Clare. I abandoned the world and met nobody. When I was alone without a family or friends for those two years, I experienced great desire – desire for shelter, desire not to be alone, desire for warm food or a soft bed – but I also learned that all desire passes, if you allow it to. I learned how to meditate and be silent, while alert to the passing world.

My father was filled with desire. He broke his own heart trying, and failing, to be a man that fitted his own definition of a man; and because I did not fit that definition, he tried to brutalise me into the toughness that he thought of as masculinity. I was tall for my age, my voice deep, but it came out crushed to a squeak when he was about. He failed to turn me into a man like him. I am weak, a wastrel, a coward, to his eyes. I have always had pretensions above my station, and it caused me many black eyes.

He strove for meaning. There is a sadness in a creature that strives for meaning in a world that has no meaning. I heard him once in the toilet, sobbing. That is when I realised that he had feelings, too. Before that, I simply thought of him as a hard man. He had always wanted to be somebody. He had never realised that he was somebody – he was my dad. For many years I thought if it didn't hurt, then it was not love.

The choice I made to leave home and start a wandering life was not mine; it was forced upon me and I was unwilling to take it, but it was a great gift. I went into the wilderness and travelled alone in nature. My life as I knew it – life as a boy – was torn away and I was ripped open, and inside there was a hard and devastating loneliness. But as I grew used to living outside, the plants and birds became my company, and I realised that what had filled me before, what had made me feel real, were thoughts about wants and desire: thoughts about myself, fears about work and money, fears about people, feelings, wishes, hopes. When those thoughts left, it felt as if there was nothing inside, and that for a while was frightening; but the fear passed, and into the nothing came a feeling of equality with all living things, with the rocks and the water and wind. I realised that I was just like them. I was part of the world, an emanation of the creativity of the earth. Overnight I became responsible for my own life; there was no television to tell me what to think about, no kitchen that might have a snack in it, no heating system, no lights to turn on or off at night. I owned the choices that I made, and I owned the consequences of those choices. There was no human, no shaman, tyrant or demons telling me what I should do, how I should behave, what was normal. I was set free like a balloon and I had to make everything up for myself.

The rhythm of walking encourages thought and, as the body wanders, so does the mind. The repetition of putting one foot in front of the other is liberating to the mind, and ideas come and go and reach destinations. There would be no return. There was nothing to return to, and so I became just and only myself in the flow of the world, free to choose my way; and when things failed, I was free to accept and embrace that fate without disappointment or anger. I existed in reality, content to live and choose and be responsible for those choices each day. Even to this day, so many years later, I feel wholly responsible for the footprints I leave and the damage that I inevitably do, that we all do by living and consuming. I came to this realisation in my walking and my emptiness, which was full, brimmed with life and potential and wonder. I realised that each flora and microbe in my body is a flower and, with the right nutrients, will flourish for its lifetime and then fade. And each body on this Earth is a flower, and the mind it holds is a flower that will flourish and fade; and we are all flowers on a planet that is a flower in a world that, with the right nutrients, will flourish for its lifetime and then fade. And everything is all right, and will be all right for ever. A Tarot Fool sniffing a bloom and dressed as a flower. And what did I care for a job or a car or a house or a holiday?

When I came back to the world I was changed, and now I cannot help but walk; it nourishes my body and my mind, and it is what I do to escape becoming part of a thing, a group, a thing with group-mind. Travelling alone, I am responsible. Long-distance walking is painful and lonely, but it was my chosen pain and loneliness. Just as the pain of my work is my choice and I am content with it – nobody tells me to do it, I do it of my free will.

In my imagination, this life has been a path with many, many forks, each one a choice to be made. Each unchosen route fading from view as it became the past, its destination unknowable. No destination is really known until you arrive, and then it becomes merely a point along the way – a vague place rarely planned for, simply the start of another adventure. The only thing to do is be happy with the outcome, whatever it is. The path leads to the end, as all paths do. I've had some rocky paths and dead ends, and decisions that led to disaster, and others that led to love and passion and poetry, to excitement and adventure. All I can do is embrace them all and move on. People sometimes get frozen and unable to decide which path to take; others instantly regret their choices, because their dreamlike fantasies about the unchosen path were far brighter in their minds than the reality and effort of their chosen one. What could have been has never been, and will never be. This is the Tree of Life where each branch grows and bears fruit and, ultimately, ends in a bud. There are no rules, and nothing planned by humans is ever planned that way again. The way is vague and unknowable.

We cannot know what the future will bring, and 'What we cannot speak about we must pass over in silence,' says Wittgenstein. When circumstances mean that I have to make plans, I keep my plans woolly, so they can change. Why try to be firm, when the world is vague and the future unknowable?

Colchicums

It has been dry for a few days and the corn poppies and field poppies, cornflowers, thistle and wild carrot are tanned and fragile, as their seed has gone and the sun has dried them out. I'll let them stand, because they are elegant and charming just as they are. They are the first visual signs of autumn, although on a cool, dewy morning the scent of autumn comes first. The butterflies too are gone, but small birds feed at the seed heads. I'll try to cut the meadow down by Halloween.

Some of the short, oval blades of cherry leaves are orange; they remind me of a time working with steel, hammering out white-hot metal into tools to carve stone with. If the leaves were glowing iron in a blacksmith's forge, they would be nearly at the heat where they could be quenched in cold oil and hardened into knives.

Out under the beech trees in the shady places, the little pink cyclamens (*Cyclamen coum*) are flowering again. They have spread over the years among the colchicums (*Colchicum autumnale*), which are also flowering now. The cyclamen flowers grow from a corm that just shows through the leaf litter. It is concave on the top and convex on the bottom, like a rough, hairy brown potato lens, a couple of inches across. The flowers grow out of the wide depression in the top on thin copper stems, and a few hairy roots grow out of the bottom. Over the years the corm grows bigger. The flower stems curl upwards; the small pink flower faces down, the petals flung back like wings, showing all its brazen workings. As the flower fades and the ovary behind it swells, the flower stalk grows longer and twists into a tarnished copper spiral spring, then when the seed pod is too heavy and about the size of a small marble, round and reddish-black, the stalk can't hold up the weight any more; it slowly reaches out and gently places the pod on the ground. It splits open, the seeds fall to be carried away by insects and, if the conditions are right, each seed will grow into another cyclamen. In this way the cyclamen keeps its family all around it in its little village, where it was happy, and its young will grow where they should be happy, too. Eventually the floor of the woodland will be carpeted with cyclamen.

The colchicums that they live with are also pink, and they look like tall crocus, fragile pink cups with delicate bare stems. Some people call them 'autumn-flowering crocus', but they are not related. They are thin, fragile tubes that break if you touch them with even the lightest finger. Yet rain and wind do not bother them.

Miss Cashmere totters out to see the dahlias – they are at their best right now. I am on my knees, snipping the soggy heads off the flowers that have faded and putting them into a bucket. I look for a bud that is about to open and I cut the stem off just above that, so the sap goes directly to it, and not beyond. The wound will quickly heal. Nature makes good scar tissue – it's the very substance of growth, filled with special cells that create a barrier of sap or pus or resin against infection, and busy active cells that grow across and close the wounds. Each dahlia that I remove will be replaced by two others. The plant's job is to produce seed, and the loss of a flower is the loss of seed, so to compensate the plant produces yet more flowers until the frost comes. That is the theory, and that is one of the reasons gardeners deadhead. But it isn't strictly true. This plant is cultivated by plantsmen in factory greenhouses; it's over-bred and sterile, but it doesn't know. If left alone, the plant would produce the flowers anyway, but the new flowers would be lost among damp and falling flowers and the garden would look uncared for. So I deadhead to make it look pretty, because that is what I'm supposed to do. Personally, I prefer the colours of fading flowers, but then I'm a godless heathen and my passion is for fertility, decay and rebirth.

A perfect orange dahlia, its globe of petals tightly curled into a million tiny tubes, makes a flower the size and shape of a golf ball that I am tempted to hold gently in my fist. Wrap my rough, grubby fingers around its papery texture, feel its lightness, feel its texture on my skin. For a moment I want to own it, and at the same moment I am of course aware that such things are not available to be owned; that moment of want, and the knowledge of loss, is both the joy and the sadness of life. In a few days the flower will be gone and will never return, but there will be 10,000 other flowers, 10,000 other moments like this. When it fades, like all the others, I'll snip it off and throw it on the compost. I reach out to hold it, just to feel its shape and weight, but it doesn't feel like enough. I am tempted to learn what it feels like to squeeze all the air out, close up the gaps between the petals; slowly crush it into a solid ball. I let it go. What we think of as love takes many forms.

As I move away, Miss Cashmere hobbles towards the dahlias. She takes one, holds it in her hand, a medium-sized one, the size of a tennis ball. Then leaves on her short, thin legs and long feet, with steps less than the length of her shoes. Looking somehow much older today, as if she has stopped trying to cling on to youth, stopped making the effort to maintain the illusion. Each step a half-step, a quarter-step compared to mine, she totters to the house, then comes out again with her scissors and a vase and snips all the orange flowers off and, with a quick lean forward and a twist, as if she has a pain in her hip, she slowly totters back inside. These four journeys take her quite a while – she is making an effort. If she had asked me, I would have done it for her. Perhaps the effort was part of her reasoning. Perhaps for a moment she was Sisyphus. I will never know.

I am surrounded by buzzing life and the sun has become warm on my back and the air stilled. Warm skin and flowers and insects surround me, and as the damp earth warms under the late sun, the scents and the sounds fill my senses. Kneeling on the earth, bathed in scent, sound and colour. A taste of no-self.

Scything the Meadow

It's time to cut down the meadow. I abandoned machines in this part of the garden a long time ago. It starts by taking the scythe blades from the oily canvas rolls they have been stored in, checking the edge by looking for cracks or chips. If I press my thumbnail against the back of the blade, the edge bends a little; it is as thin as paper, razor-sharp. There are a couple of chips where the metal has broken away, perhaps after hitting a sapling. I file into the steel to make the edges smooth and then, sitting on a stump, I place the edge of the blade on a little anvil and I hammer the metal out, drawing it from the back of the blade into the curve. This process is called 'peening'. Slowly the cold metal stretches out, the blue steel hammered to a thin edge. I tap all along the side of the blade and make the edge a razor. The blade gets peened and adjusted before every use. It's like a tennis player checking his racket, a hunter oiling his gun, an archer setting up his bow. With a whetstone as smooth as skin, I swipe along both sides to take off any burrs or microscopic bits of crystal hanging on it, which could catch and start a chip or tear. Overlapping strokes that clang as the stone hits, then swishes along the ringing blade to hone and polish the edge.

With the blade in my left hand, I clamp it to the handle – the 'snath' – and adjust it, so that its angle to the ground is suitable for my height, my arm length. The back of the blade needs to slide flat against the earth, without digging in at the tip or along the edge, so I adjust the angle, with a shim in between the attachment tang on the blade and the wooden snath, and clamp it up tight. I know that a two-pound coin is the right thickness of shim for me, because years ago I measured everything, adjusting the handles and the blade angle, and tuned it all to suit my body. I have not grown, and the snath has not shrunk, although I am a little more bent than I was – and by the end of a day of scything even more bent – but the two-pound coin still seems to be about right.

Perfection is never achieved. There are other adjustments to make: the tip needs to be on the same radius as the heel or 'beard' of the blade, so that it catches the grass and throws it to the side. A poorly adjusted blade can break the light wooden snath it is attached to, so I take my time. All set up, I spin it round to get the feel of the scythe, swing it around my head. It feels good to hold it again; it is light and balanced and, although it is hot, I am looking forward to using it. The blade flashes in the sun; it is three feet long and sharp enough to cut the feet off anybody who would step into my path. I celebrate this tool that I love perhaps more than any other, not used for months. I want to get the feel of it again. I fill my stone holder with water from the standpipe by the greenhouse, and hang it on my belt. In the past, the scythe-stone holder was often a cow horn, with a metal hook to hang it from; mine is made of tin. The whetstone is a smooth, oval piece of limestone about eight inches long. It goes into the holder and, after every few strokes, I wipe the blade with a lump of grass and stroke the stone along the edge to take off any burrs.

The stalks are dewy and they cut easily. As they dry out during the day, they become more difficult to cut, so I try to do this job early. The earth is rough. I go slow and low, to leave as little stubble as possible. The meadow-flower stalks are long and fall off the blade at the end of each stroke, leaving a windrow to my left that will dry in the sun for a day or two. A hedgehog appears out of the grass in front of me, and I put down my scythe and move it out of the way with my foot. A lying unmanned scythe is a very dangerous thing, and I go round it carefully to pick it up; once in my hands, it is safe again, even-tempered and benign, and it loves me back.

I become lost in the rhythm of the cut, and the whisper of the blade slicing through the stalks as they fall and tumble in a row four feet away. I stretch out to my right with hips and arms, to cup the dewy stalks in the blade's curve and hold them briefly, lovingly, in the arc of the now shiny-wet blade, before throwing the whole thing away to my left with fully extended arms and waist. The blade flies close to the ground and slices through everything and throws it off the blade at the end of the swing. Then I swing the empty blade back as far to my right as I can, and take another short step forward to recharge the whole machine, and swing again and cut at full extent a clear swathe more than seven feet wide.

Scything is an art that has been lost and reclaimed. Celebrated by Robert Frost and Leo Tolstoy, who wrote about its exercise and beauty, it is a peasant's job that is envied by the privileged. Small farmers in the UK are taking it up again. Council workers in France and Germany use scythes to cut grass verges. I love it because although it is physically hard work, it is a poetic way to spend a couple of days, and everything sings around me as I work.

It is hot and I stop from time to time to listen to the birds and hone the blade. Scything needs no protective clothing, so I wear shorts and light boots. Miss Cashmere is away, so my shirt is hanging on a tree by my lunch box and water bottle. It takes me three mornings to cut down the whole meadow, which looks bare and sad when I have finished, like a house after the Christmas decorations come down. If I didn't cut it, it would in a year or two become grassland with a few wildflowers; the cutting helps to protect the slower-growing flowers from being over-whelmed. Although I love the scythe, if it were my meadow, I would let that happen.

Autumn Equinox

We sit at the tip of an hour hand, slowly rotating and watching the days pass by. Today is the equinox, the first day of autumn. One of the four cardinal points in the Earth's cycle. This equinox usually occurs around 23 September; it can be a day or so either way. On this day the sun rises directly in the east and sets directly in the west, and the nights become longer than the days. The light is reduced by three minutes each day, but the rate slows down towards the winter solstice around 22 December; then the day is at its shortest, and we move on towards the spring equinox around 20 March, when the day and the night are of equal length, growing longer until the summer solstice – the longest day of the year – which happens around 23 June. The cycle goes on, round and round, and everything becomes a little older. For people who work the earth, each of these points marked the first day of the season and a time for celebration, because farming crops and the husbandry of animals and the security of the family are all about day-length and temperature. Many of us still mark and celebrate these older festivals.

The weather forecast will tell you that the seasons begin on a different day. Meteorologists use a system based on average temperatures, rather than the cycles of the Earth and the sun, and for them the autumn is simply September, October and November.

All these points are roughly ninety days apart: a quarter of a circle. The dates of these events vary from year to year, because the Gregorian calendar defines a year as 365 days, but it takes the Earth slightly longer to complete its orbit around the sun. This means that the autumn equinox occurs about six hours later every year, and this eventually moves the date by a day every four years, hence the leap year. During the twentieth century it was also found that the Earth's rotation was slowing down – days becoming 1.4 milliseconds longer each century. All measuring systems are, by their very nature, slightly imprecise; no matter how fine our tools become, there is always a relative-sized gap between the fit of the tool and reality. The inaccuracies become finer, but are always there.

The autumn equinox is a time of change – a time when old things end and new things begin. In the garden the living and the dead both exist at the same time, side by side. In all their stages, seed and flower and dust, there is decay and birth; it is neither today nor tomorrow, and it is both.

It has been a long, hot summer and the farmers are driving round the fields in tractors, baling straw. The crows in groups are flying high across the aeroplane trails and the pinking evening clouds. The days are growing shorter by three minutes each day, and the blackbird no longer sings. Time to grow the winter beard, find the woolly clothes, gather in apples, tend the log pile and put a good log aside to decorate for Yule. It feels like travelling home to a warm hearth.

October

Go Now, Bonnie Boy

The moon's bowl tips a group of crows in pairs at dusk, and I am showered with gifts of rain. Seagulls scream in madness at their loss of seaside home. They have gone quite mad. They think my roof is a clifftop, the dustbin cart is a trawler, the black plastic bin bags they rip open seem to them to be the carcasses of seals; they throw the plastic bones across the road and pick out the sweet entrails: chicken skins and burger remains and pizza crusts.

Laughing couples trip into the Forester's Arms to dance, and I alone, as I'm alone, stay out and head for home with whisky, and remember when I was a wandering nature boy who didn't want a house, or a job or the chains those dread things bring, but in the end I allowed them to accumulate, for my children's fitting in. I remember then seeing men who were the type of men who wore shiny shoes with patterns of holes punched in the toes and polished into newly-peeled waxy chestnut gloss. Trousers and jackets. Shirts and ties. Knitwear. I remember wanting knitwear and shoes like that, and I am aware that now I am just like them, and I have shoes like that and a house.

In the car park under the machine, a boy in broken trainers asked for change, some coins. A child balanced on the thin razor edge of existence, which could carelessly slice either this way or that, depending on the amount and quality of the gravity that tugs him. I emptied my wallet and gave the notes to him, trying to add some weight to the side that meant life, aware at the same time that more life meant more suffering. I kept the whisky, which he probably needed more than I did, and I clopped away in my shiny chestnut, polished brogues. As I left, the council van-man, yellow jacket, came and moved him on.

'Fuck off!' he shouted from the safety of his van at the boy, who picked up his bits and pieces, his tattered little child's rucksack and his filthy sleeping bag and moved on, to die perhaps or just to be somewhere less shameful, for the people who pay the wages and buy the van and yellow gilet. I put my ticket in the slot and drove my van away and thought: Go now, bonnie boy, go to the country – the country can fix you, I cannot. Leave this cold, uncaring island if you can; go to sea.

October Mist

The ivy is flowering in great mace balls and is crowded with bees. Toadstools grow in the lawns, their white threads of mycelium feeding on the decaying roots of long-gone trees. The soft heat of fading summer rallies for a while, and winter is in remission. There are still strawberries on the plants fattened by the rain; those touching the earth have been nibbled by slugs, but those higher up are plump and bright. I pick a couple and eat them, sweet and juicy and red, like my hands and cheeks. I take a basket of them up to the house for Miss Cashmere, leaving it on the picnic table outside the kitchen door. She is nowhere to be seen. Her morning routine has faded; years ago she used to be up after the children went to school, and I would watch her go down to the summerhouse in all weathers with her newspaper and cigarettes, stopping for a chat sometimes. Always with her newspaper.

Under an apple tree that is heavy with fruit and is casually dropping big cooking apples, the blue trumpets of gentians (*Gentiana*) are happily blooming in the semi-shade among the fallen leaves and Bramleys. There are hundreds of varieties of this famous medicinal plant. Where there are leaves and leaf mould, the soil is acid and these little acid-loving plants are happy, their trumpets pointing up and out against the decay. In the borders the *Hydrangea paniculata*'s massive white flowers have now mostly turned pink and are starting to go green, as they do every year; then they will dry and crisp and catch the frost, and in the spring I will cut them off, down to a good pair of buds, so that it can do the same thing all over again next year. I'll take a bunch of dried pinkish-green and blousy flower heads home and put them in the vase on the sideboard, where they will sit until Christmas, when I will replace them with holly and red dogwood and ivy.

The quinces are heavy by the wall, the leaves gone, and the fruits tucked in their protective cages of thorns are big and green. In the early spring this hardy shrub makes big red waxy flowers, and then I ignore it all year until I pick the fruit, take it home and make quince jelly and quince cheese, in an almost industrial operation that takes three days of soaking and grinding and cooking, and then another day of cleaning up the staining rose-red mess. The quinces are mostly green, and their stone-hard flesh is pale green to white, but when they are cut and soaked in water, the liquid and the flesh turn pink and make the sweetest jelly to eat with cheese. I still have big jars from last year, so I won't bother this year. I'll let them fall for the birds.

There are petals on the ground and the garden is green and pink. The leaves of the cherry trees that I planted so many years ago are turning a bright raspberry; the crab apples shine yellow against the dark branches and are attracting blackbirds. The acer has gone red, as the chlorophyll and the sugars are sucked back into the tree to keep it alive over its winter sleep. The roses, too, glow red through the October mist. The last few roses. Apples, dahlias, raspberries are all bright and clean and red. Walking the dark earth, I am a peasant eating its bitter-sweet fruit.

After two days of unhelpful weather I cut the grass, perhaps for the last time this year. Up and down, turning my back on the sun, then a long, slow walk towards it, head down to keep the sun from my eyes. My shadow lengthens, the sun is low and I appear to be tall again, but curled. The years drift by, my shadow shortens and I start to bend. A sadness washes me again as the leaves turn gold. A sadness that always comes in October. I've heard many a gardener say the same: it's part of the trade. The garden is muted by a soft drizzle and a greyness that fills me with darkening feelings. The mornings and evenings are cold now and I need layers, but the sunny afternoons are hot. I pull a fist of flowers from today's soaking garden to take home – marigolds for Peggy.

Birthday

This month is my birthday. Should we mark these events or not? I like to get presents. To feel special, to be somebody's flower. 'Tell me it is just a number,' a friend said a few months back, after his birthday. So I searched 'It's just a number' and the search engine gave me 4,650,000,000 results in 0.37 seconds. Numbers so vast and so tiny as to have no meaning. My mind is inadequate to deal with numbers – especially numbers too large to describe everyday life. I do not trust numbers, beyond those I can take in and count in a single glance: eight or maybe nine. Beyond that, they stop being individuals and become a larger or smaller group. Quickly I get tired of looking at page after page of questions from people fearing the passing of the years.

Words are shaped to fit our mouths and our culture, but they limit the ways we are able to see and think about the world. We can describe it only in the language we have available to us, and therefore we can only think about it in those ways. I want a new language: a language that enables new thoughts rather than regurgitating all the old ones, a language of the smaller things. I want a word that describes the FreedomLaughterPeaceStrengthColour of ageing. One that describes the ConfidentCarefreePainJoyBeauty SadnessLiberation of these conflicting feelings of Fearless-Fear that joyously interact with each other and create a layered complexity that is kaleidoscopic and constantly evolving. A simple word. Perhaps the word 'mature' comes closest, but I play with immaturity constantly, and Peggy and I are in fits of laughter pretty much every day; I hold her tight and ask what it's like to be fondled by a pensioner, and she runs off screaming, shuddering and laughing.

I used to wear a watch at work, but lost it when I was cutting a hedge. Two years later it was found by one of Miss Cashmere's children and I got it back; it had counted the minutes and hours and days while nobody was looking until it had wound down and stopped. It said it was Tuesday at 11.00 when I picked it up on Friday at three o'clock. I wound it back up and it started counting again, now it runs a little fast, gains a couple of minutes every day. Dumb machine knows nothing about time. We living things are time itself – opening then closing, like the flowers at night. Pacing time through our changing lives.

Nature does not count. It only knows 'need' and 'not need'. Man has a childish relationship with numbers – 'How much have you got?', 'How much have I got?' – whether it is money, time or love; and 'What does that say about me in relation to you?' Nearly every time I go into the Forester's Arms there is a man, who I think is called Martin, and his wife; he has a haulage business. I did know his name, but have forgotten it. He gets drunk a lot and is very unhappy; every night he destroys a few more cells in his brain and his liver and kidneys. His wife gets drunk, too, and for the same reason. It does not matter how many houses he has got, how many employees, how expensive his new car is – he can't make himself happy with the numbers. He tells anybody who will listen about his car or house or holiday. He thought the numbers would save him. He has lost his connection and is trying to destroy himself.

How can I explain the joy of being nothing, of having nothing, to one who wants things so badly he is willing to destroy himself, annihilate his peace of mind for them? There are much happier ways of destroying yourself. Over the years I have been getting rid of stuff; there are fewer things in my life now, and I enjoy each of them far more. I think this is what Miss Cashmere is doing by getting rid of her books. My birthday is coming and I feel it should be important, but it is not, yet I still think about it. How much time have I got left? What shall I do with it? This autumn is so far the best time of my life.

Whisky

Drinking a sweet, dark, smoky whisky with the door open to the cold and mackerel sky, and reading Robert Frost, who talks of simple things. A butterfly, scything a meadow, a small bunch of flowers, and these things fill me and drive out the cynical. He reminds me that all is the void, and we look where we choose to look and see what we choose to see. This evening I have chosen whisky, birdsong and the different sounds of drifting leaves.

I can smell the coming winter. The falling heat and rising cold. When it's here, I'll feel it on my skin and, when I do, I'll remember the spring when the grasses grow and daffodils open; and when that happens, I'll be reminded of summer and I'll want to feel the heat and see the insects and the birds that cruise and herd and eat them while I'm sitting indolent against a tree and searching for coolness. And as I sit in the heat, I'll be looking for the leaves to turn to gold and drop and bring their massed colours of decay, and I will smell the coming autumn. And at each stage I will fight the immense temptation to think ahead to the future, and for a short while I will fail. But now I smell the winter and wait for cold to make the grass sparkle, and for the air to be clear and slant; and I will be able to hear the freight trains pulling hard over a mile away and the owls a quarter-mile away, and the foxes yapping and cats crying like babies, as the cold inverted air keeps the sound in and bounces it round to my hungry, devouring ears. So each morning, every evening, I look out for the first sparkle of frost and, when it appears, I'll feel that flood of excitement, that deep, visceral gut-tugging twist that a young man feels when sitting with somebody whose eyes have just connected with his and he realises that their bodies will still be together when the sun rises. Because I love the cold.

I finish deadheading the roses this month. From October, I leave whatever blooms remain on the stems and they'll grow brown and mushy and fall to the ground, and from their darkness and shadow the green rosehips will grow fat and ripe, and become red fruits that feed the birds and squirrels and add a flash of colour in the greyness of winter. I hope the snow falls and rests on them and their thorny branches, or frost grows along the twigs.

In the light, almost non-existent rain the sparrows sound delighted. Invisible until they move in and out of the massed red berries of their safe space, in the vicious firethorn by the gate, whose spines pierce the thickest gloves and burn like fire in the skin. Pyracantha and, by them, flabby white snowberries, fat and lardy-looking. The acers drop the first of their leaves. There were none on the ground last week, but a few scatterings below the trees shine and catch the eye like banknotes. A few hundred quid's worth. The red ones have gone red and the yellow ones yellow, and the liquidambar tree four times taller is fading from green to gold, as daytime darkness creeps in and the velvet shade that has hidden in the corners all summer long begins to become dominant, restorative, breaking down and building up and healing the exhausted plants and soil, and the souls of those who care to look.

Where Miss Cashmere can see them from her kitchen, at the top of the wide stone steps that go down to the garden, the pelargoniums in the pots are still red and flowering, but looking a bit straggly. I cut them back, put them in the greenhouse and, if we are lucky, they will come again next year. The hydrangea by the door is still cream, but turning green and pink. And through the dried-out stems of fennel in the little herb garden, which she can see from her window, the orange nasturtiums have fresh green leaves that defy the season. Marigolds still flower on stray grey stems.

The horse-chestnuts are all down, thousands of them, and make walking difficult. If she were to step on them, she could twist one of her thin ankles and fall amongst them. I still wonder, after all these years, what is the best way of picking them up off the drive and the grass. I wander over to the stable where the old tractor lives, which we ('we' – I still say 'we') use for mowing the bottom garden, lower the cutting blades and drive round sucking up the conkers. The ones that have been planted in the lawn by the squirrels will sprout in the spring, and their pairs of bright virgin leaves will be cruelly cut off with the grass in March by the passing mower. The red-berried rowan sings so loud, with its massed flock of sparrows as I drive by, that I can hear them over the little diesel engine.

Molecatcher

Peggy and I went to the pub last night and we met Byron in there. He asked me what I did in the winter and I said that I used to catch moles, but this winter I'll be working on another book (this one). He told me about his brother, who was a miner, and during the miners' strike he had nothing to do and no way of making any money, so he went about catching moles for farmers. He was very depressed, Byron said. Molecatching is an isolating and lonely business. Molemen wander over distant fields miles away from anywhere, places where only cattle and sheep, horses, crows and molecatchers go. One day his brother didn't come back. Byron and the rest of the village thought he had killed himself. Byron told us that he and his dad, and some other men from the village, wandered around for weeks looking for him, but he was never found.

I could have found him. I know where these old vagrants like to hide, to spend the odd night. I can sniff these places out – it is in my blood. He will be in a wild, unvisited gap in a hedge, between two rows of old woodland trees. In that gap there are good resting places. He will have hanged himself like a farmer, from baler twine, or taken rat poison like a gardener. His eyes would have been had by crows within a day; it is good that they didn't find him. I've watched the crows pecking the eyes from feeble animals: newborn sheep, even before they can stand.

Nature has no mercy. His bones will not be found until the hedge is dug up, unless a fox drags some body part into the field and leaves it there while a tractor goes by. Nobody tends those old hedges any more; they are cut from high up in a tractor cab, with a spinning flail that just whacks the tops off. The drivers don't look down.

That evening the horror that Peggy and I will one day be parted creeps up on me intensely as I lie on my bed and read of Elizabeth Barrett Browning and her love of the sea, which took her brother as its own, but I keep the feeling to myself. Peggy and I would both end up weeping if I were to say anything. I choke it back and get up, and go to sit next to her while she watches television. I hold her hand, squeeze it tight, and she turns and smiles at me but says nothing, as if she already knows.

I love the feeling when the old night rolls over and goes to sleep and the fresh morning creeps in: both are weak and in balance, as if the night could come back and the day fade away. Things at the edges of being. At dawn I like to leave my bed and watch and write in my notebook, and then, if I can, go back to bed for an hour and feel the tingling drop of almost falling asleep, but not quite making it. I wonder why I love the marginal things so much, the tender things that might not be there, or could be gone so easily: my old brown mug, the fading leaf, the old lady, the newborn seedling, the invisible line between awake and asleep, life and death. And it dawns on me that this is where life is exciting – on the margin, at the crystalline, hammered-out knife-edge where anything could happen.

The summer has passed, my busy work done. I am slipping back into old and comfortable ways: waking at four, then rising at six to watch the clouds brighten and hear the birds wake each other up, and feel the cold on my skin and become contented again as the autumn sadness fades away. It is dark; the sleeping people's mouths seem stuffed with feathers, the light is muffled. I am waiting for the day to start properly.

The basket of strawberries is still there on the picnic table. They are firm in the cold. A blackbird perhaps, or a starling, has been pecking at them.

The grass is wet in the mornings and there are toadstools of various kinds springing up in the lawn and by the paths, and on the dead and dying branches of trees and stumps. Each day a new crop of bright-yellow or small brown fairy parasols in tiny bunches has sprouted up overnight. In the middle of the lawn a mass of brown meaty clumps that weren't there yesterday. The luscious woody smell of mycelium. The slugs love to eat some of the toadstools, and over the next three days some of the smaller, brighter ones remain firm and the bigger, darker ones fall to pieces. By the edges of the gravel driveway there are shaggy ink caps – edible, but they soon digest themselves and turn into black ink within hours of picking. I used to make ink from them, and from oak galls, when I was learning to paint. I would sit at my table for hours mixing ink and trying to paint circles – freehand circles over and over again – trying to draw them as perfectly as I could. After painting and drawing thousands of them, some slow and deliberate and others fast and free, I was able to draw an almost perfect circle, freehand. The most perfect of the circles seemed to me to be dead things and I found beauty in the imperfections. My tutors often complained that I could control my pen or brush and draw a perfect circle, but my handwriting was an illegible, immature scrawl that I needed to translate for them.

I know a couple of the edible fungi and I like to look for oyster fungus, which often grows on dying beech trees, but I have found none for the past two years. The public woodlands that I sometimes visit are kept tidy, and dying trees are cut down, perhaps so they don't fall on people. The beech trees in this garden are too healthy and I look at the woodpile in hope, but there is only candlesnuff fungus, which looks like little flat, dark-brown antlers with white tips, and some wavy-edged disc fungus whose name I do not know.

It is one of those quiet days, as if the world is still hiding under the duvet, a day that never really gets going; even the birds are silent, the air is still, the milky sky is full of no colour and I am adrift on a sea of rolling green, sandwiched in between and, seemingly, the only moving creature in a silent world. Feet connected to the earth and head to the sky. I have not heard a human voice since Peggy's last night, and I have not used my own, and slowly the day is drawing to a close.

Miss Cashmere dodders past on her thin ankles and long shoes. Smoking a cigarette, as usual, carrying her pack with her; the sleeves of her old cardigan stick out of her coat sleeves and come over her hands. She has a decayed prettiness. Like a dusty old jar of faded face-powder found on a dressing table in a museum-room display. Her thick brown tights or stockings, her tartan wool coat. She doesn't notice me or, if she does, she does not acknowledge me. In the sky there are ripples like those left on a beach after the tide has gone out. The tide comes in, the tide goes out; its regularity could become monotonous and yet it becomes stability, becomes the only living, breathing thing.

Our Lady of the Flowers

A fist of flowers stolen from yesterday's garden, just for you: marigolds, a few roses, that's all that I could find worth collecting. I put them in an old clay vase and stand them on the sideboard.

In the flickering, fading evening light Peggy, my ladybird, is unreasonably sad beside me on the sofa. She says that she does not know why. She opens wine, and I accompany the rain's white drone with clumsy Irish jigs on my mandolin. Cat sits at my feet. I'm playing gentle trills, triplets and rolls, and trying to make them sweet; an egg boils on the stove, playing little rattling notes. The cat climbs up, curls, leans against me and dozes.

We rarely disagree these days. If one gets the better of the other, does that mean that one is right and the other wrong? I am no expert on anything – I have no wish to be. I turn on the radio and I hear people who consider themselves experts disagreeing and invalidating each other's arguments. The strength of their opposing beliefs enlightens nobody. I turn them off again. Is this the kind of thing Miss Cashmere reads in her papers every day?

The year is ending, and I deal with the autumn melancholy by reading. I say, 'I love you, baby' to her and she smiles and says, 'I know.' The words 'I love you' are not adequate; I think perhaps because there is an 'I' in the sentence. I'm drinking hot chocolate and reading Patti

Smith's *M Train*, which is all about coffee, at the point I'm reading it. I'm trying to read with blurry eyes without my glasses, which I am too lazy to look for, so the lines keep skipping and words merge, and I get a whole new non-linear interpretation of the words; and a dream-like state comes over me, of coffee beans and hotels and plaster saints and camellias, and slowly I begin to fall asleep, while Patti Smith whispers of Jean Genet and I remember reading *Our Lady of the Flowers*. As I drift off, I think of the flowers that close at night and open again during the day: morning glory, osteospermum, tulips and Californian poppies, then the hibiscus, which closes at night and never opens again.

Apples

The leaves on the cherry are much pinker now, and in the park the fallen horse-chestnuts have been collected in carrier bags by parents and children, who just want to take them home because they are shiny and pretty and plentiful. The cow parsley's brown stalks in the bright-green grass are heavy with seeds and stand tall and wait for the wind to snap and flatten them, the winds lifting their little fat flying-saucer seeds and flying them eastwards. Each year the seedlings appear closer to the woodland, but will grow no further because they are sun-loving plants. The holly berries are bright and waiting for Christmas. The crab-apple leaves are drying but hanging on; there's yellow fruit, and a blackbird feeding on it chatters an alarm call as I pass. In town I bought myself a little bunch of freesias to replace the wilting marigolds.

In the garden I pick the last apples from ancient trees. The trees are diseased and cankered, as old things are. But the apples are crisp and their cider will make you feel like a king. Perhaps I'll make some again this year. Like the trees, I am older, but still churning and still hungry to be bitten. Years seem to fall like snow now, piling up like leaves that blanket the ground like a silent duvet.

First Snow

Today the first snow for just a minute, then hail bouncing on the shed's tin roof, then sun, wind and clear sky for the rest of the day as the trees bent and flicked. The passers-by are also bent, wrapped in wool and fleece going about their day, and I decide to stay indoors and watch. The watching stops and I stop being for a while, then slowly come back and much of the day has passed. I have become a cat, or a sleeping dog that should be left to lie; but more a cat. I used to work at being and not being, doing and not doing. Watching and waiting, but I realise that I no longer wait; I haven't waited for anything in years, and now watching simply means watching. There is nothing I want or need to do. I have no plan. I do not hope. I do not wait, I do not wish, I do not fear. I do not be or not be, do or not do. I just seek comfort, avoid discomfort, curl up in a warm spot and watch. I have become a cat and have reached the peak of my evolution.

I sit on a cushion at the open back door. My breath comes in, my breath goes out. There's wood-smoke in the air and the smell of leaves, and the smell of fox and cooking, and somebody's television game show: orchestrated laughter and clapping. The light goes, clouds cross the stars. It's getting colder. It is too cold to sit here for much longer. I've spent all day wrapped in a blanket and reading

Edward Thomas from time to time. I leave him by the back door, eat and get dressed, go out in the dark to listen.

A hundred crows play around the wind-bent poplars and it is so right, so perfect, that suddenly I feel like crying. Crowds of shadow-birds looking melancholy against the last bright blue. Slate tiles and tilting stone chimney stacks, flues leaking clouds of grey-black wood-smoke from newly lit fires that drift into the darkening October sky. There goes the light. The magpie has left his sentinel post on the television aerial. It won't be properly light again until after nine tomorrow, and I turn back to the house, watch the darkness become complete. Pull the cork from a bottle of old whisky and pour a glass, the scent of which keeps me occupied for hours.

November

Hop-tu-Naa

Porridge for breakfast with a tot of smoky whisky. Thick woollen socks. The poplar trees are bending, looking pencil-drawn and scribbled. Last night's sketchy crows appear again, dozens of them suddenly, as if they were birds just of winter and wind – a favourite creature of the poets, the horror writers, the goths, the Celts, the Native Americans, the bereaved. The crow has power over all who see it. They are us and we are them, and we see meaning in their movements when there is none.

Last night and today until midnight is Samhain, Halloween, in Welsh *Nos Calan Gaeaf*, which means 'the first night of winter'; *Hop-tu-Naa* in the Isle of Man where my grandmother, who gave me a home, was born. This Manx word seems to be the origin of the Scottish Gaelic word 'Hogmanay' and was the original Celtic new year. Harvest is in and winter is on the doorstep, and we are prepared for its bite. Children would do as I did: hollow out a turnip, pale and green and as hard as oak. My father would sometimes help us get started by using an electric drill, if he wasn't in a foul mood. Using sharp kitchen knives, we would sprain wrists and bruise palms hacking out the middle of the turnip. It was only possible to carve straight lines into it with a kitchen knife, so we would stab out a rudimentary gap-toothed mouth and triangular eyes, then find a glowing ember from the coal fire and drop it inside. The smell of burning turnip would hang around the house for days.

Halloween – *Oíche Shamhna* in Ireland, where the tradition originated perhaps. My people came from these places and gave me these painful traditions. I tried to pass them on to my children, but school and television presented them with a softer, milder vision and a fear that is playful and pretend; pumpkins that can be carved by babies with plastic spoons. Those Celtic traditions have died in my bloodline, with me. My Manx grandmother's fear of ghosts was as real as it could possibly be and she, who had seen war and death, would tremble in her bed at night for fear of those who, in her world, really came back to walk the Earth and visit their old haunts.

The cotoneaster tree is dripping with red berries. There's mist on the fields, a smell of wood-smoke in the air, low sun, long shadows. I am out in the cold, beard, boots, check shirt, watch cap and chainsaw. I'm that kind of man today. There's a breeze; poplars that once were spires are now whips, they shed their leaves in whispers, to the right. Starlings draw their smoky clouds above them. Early night draws tighter and I'm cutting logs into stove lengths. I don't like the noise of the chainsaw, but there is not another way available to me. They must be cut. Later, when we get a dry sunny day, I will swing a splitting axe and stack them ready to burn down the autumn sadness and turn it into warm ash. The wind blows the smell of the saw and the wood away, it rattles the oak, ripples the grass and my beard, the hairs on my arms. We are all equal out here.

Frost

First frost of the year nipping at the last roses. Autumn's muted shades. The compost heaps are steaming on this cold morning. I can't resist touching to feel the deep warmth spreading out from the centre. A sleeping animal as big as a horse. A dragon. Distant trees glow through dawn's mist. I am honouring a great old tree by removing the ivy that wraps around and strangles it, heavy and wet twisted ropes that anchor this ancient oak to the dark, wet earth as if to stop it floating away. Rigging on a ship's mast in full sail and tugging us along. My head as high and empty as a crow's nest in the wind. Distant from even myself. I want to be a bird. A cat watches me, and I watch the cat with such an empty intensity that I become the cat, live in its body, in its mind, empty and stalking a ripple in the grass. It all passes.

This rope of ivy that I pull down could be a hand-fasting ribbon, wedding me to the earth and sweet air. I wrap it around my wrist and pull it hard, my bondage only half-complete. I am bound with vine, leaves are falling on my face and neck as I look up and tug down the twisted rough stems; a blackbird sings his last song of the day, then flies away. Who would weave a blanket of this green stuff to protect me, make it shine to reflect the sun and catch the eye and hide my sleeping shape beneath; spin a cocoon to wrap me in, where I can rest and, without a plan or thought, be reborn in spring as something else? The trees wave their arms, the roses wave their arms, the grasses wave their arms. Each man has his own God and I am tied to this one that doesn't care – the one I found after looking in all the places I could find.

My search for meaning began when I was about eighteen, after spending two years wandering the Earth and sleeping under hedges like a beetle. I was sitting on a railway bridge waiting for a train to come, so that I could jump underneath it and have it kill me. My few possessions on the granite parapet next to me: two keys, an empty wallet. I had a motorcycle, the leather jacket, jeans and boots I wore and a cold empty room; no other human being who I had any kind of worthwhile relationship with. I was alone and had been brought up to consider myself as stupid as a brick, and just as commonplace. Unloveable. I asked myself, 'Why live – why not die?' Why cling on to a life that was clearly not worth clinging on to. I sat there for perhaps an hour, no train came, and in that time I changed my mind. There was nothing that mattered and I could, I realised, do whatever I wanted. I understood then what it meant to be free. Once I had accepted death as a valid choice, I had to think about things like 'Why do it now? Why not wait a few days or weeks, go off on an adventure and see how bad things get?' Killing myself could wait a while, could always remain an option. If I ended up in prison, that would be an experience; it would be warm, there would be books perhaps and food. If it was really bad, I could kill myself there and then, instead of here and now, and meanwhile I would have had some new experiences, I would have learned something else about the world. I might even starve or freeze to death at some point, and experience leaving this place for another in that incredible way, so why do it now and deliberately?

Sitting on the cold stone and drifting away, feeling myself floating above and looking down, I thought myself warm and noticed my mind saying these things: 'Hang on a bit', 'Why now?', 'It could get worse', 'You are free to do what you want', and so on. I watched my thoughts and wondered why they were doing this – why something in me seemed to want to live. What for? There was nothing in the world I cared about, and nothing in the world cared about me. I slid down off the wall, put my things back in my pocket, got on my motorcycle and went looking for the meaning. I thought of giving up as a valid but final choice; continued life was a gamble that might or might not pay off. At the very least, I would have another adventure. I felt liberated by these ideas. Liberated from any rules whatsoever.

I had been homeless in the past, but now I was not. I had a job working on the railway, I had a flat and a motorcycle, a pile of books. That was pretty much everything. I left my job and went back to school. I sat in an unheated flat in Manchester, working in a chicken shop at night, preferring to spend my money on books rather than heat or food. I peered into the darkness to look for a lighthouse, and I found dozens of them. Too many. I spent my weekdays at college studying art and writing poetry, and my weekends in Grassroots bookshop in Manchester, reading politics and psychology and philosophy and poetry and religion and mysticism.

Politics told me that I was oppressed and, without even being aware of it, I oppressed others; the answer was to revolt – violently, if necessary. Zen Buddhism told me that life was an illusion and asked me, 'What is the meaning of a cat or a flower?' It taught me that words can only ever point the mind towards the truth and must not be mistaken for the truth itself; that words were just a map, and the map is not the same as the landscape, and yet it reminded me of the wild freedom I felt while walking, homeless, in the wilderness. Religion seemed sure of itself and I mistrusted its certainty; it told me that it would all be so much better for me after I was dead, but only if I joined the right group and followed the right set of rules and looked up to the right people. Mysticism was satisfyingly vague and told me that I really had no choices and it was all more or less preordained, but if I did the right rituals in the right way I could swing subatomic quantum things a little in my favour. The abstruse abstractions of philosophy told me that there was no certainty, no 'things' perhaps, only perceptions of things in minds – whatever and wherever and however many of them they were; and that those meaningless perceptions were perhaps all that existed, or perhaps not. Science reduced the hard 'facts' of things I could touch – trees and people – to the poetry of an infinite number of tiny spinning universes, with planets circling around that might also be waves instead of things. Psychology taught me that I could choose my perception of things, but the things themselves wouldn't change; that happiness was a choice I could make or choose not to make. Poetry told me it was all just a glorious, chaotic explosion of life and death and stuff, and all I could do was worry about it, hate it or enjoy it, and that was my choice.

I found that asking 'What is the meaning of life?' bore no fruit. It twisted in serpentine ways around labyrinthine answers. This tempting apple of a question revealed itself to have no core, no seed. It is like asking, 'What is the colour of five or the sound of one hand clapping?' There is no logical answer, no knowing; there were only ideas in people's heads whose ends did not join up. I learned that certainty about anything is a fantasy that belongs to those who need it, because their sense of self depends on it. I found that searching for meaning would get me into the hands of those who would like to have power over me, or wanted another mindless soldier to fight their cause. I wriggled out of the clutches of the Moonies and the Christians; dabbled, then veered away from, the tempting mysteries of the Rosicrucians and the Chaos Magicians; and, after tasting their bitter poison, I moved myself apart from the anarchists, the socialists, the capitalists and the revolutionaries. There were those who desired hierarchies and wanted to be my masters and teachers, and to fill my head until I had no space for myself, but only for them;

and as soon as I felt myself disappearing into their worlds, I left and wandered off to be by myself again.

I did not want to stay where I was. I had freedom, I had nothing to live or die for. I was as fixed to any square of land as the wind and just as alone, so I went looking for experience. Not wild, massive adventures. I was poor and could not afford to travel. I was tired of sleeping rough and did not want to do that again, but I could immerse myself and allow my mind to wander. I had no one to share the adventure with, but I could enjoy a deep experience of what was around me; I could not escape it, but I could explore it. Learn about the things and ideas, their relationships and connections. To look deeply at the people, to read books, to watch the clouds pass, the rain come and go, the wind move the litter; to watch the solitary ownerless cats, like me, slink by and experience what it felt like to have these things in my world and have myself exist in theirs.

Many years later, I am happy to accept that I don't know very much. I don't care enough about the details of many of the written works of men to spend any time thinking about them; they are simply opinions and, like the facts of a railway timetable, are subject to alteration and amendment. Only the poets have the truth. The world of rivers and rocks and waving grass is far older and wiser and enduring. I like my head to be clean and empty, to ring like a bell, to echo like a cave with the sound of breaking waves and wind, falling rocks and birdsong; and, like Gandhi, don't want anybody stamping through it with their dirty boots. I realised that I was not searching for an answer, but for a question.

I eventually found a worthwhile question and it was this: 'What am I supposed to do with this life that I have chosen to hang on to, this path I have decided to follow?' I allowed this question into my empty head as I wandered, and the answers came, exploding at me from all directions like clouds of seeds from a field of dandelions: 'Experience it,' said the water crashing over the rocks. 'Enjoy it,' said the crows. 'Just live it,' called the wood pigeons. 'Survive it,' barked the fox. 'Feel the flow,' said the seeds carried on the breeze. 'For what else is there?' said the clouds. This answer, as always, led to another question: 'How to bear it – how to be happy when there is so much suffering in the world?' I see the homeless and the poor and the bereaved and the abused. I was one of them, so I know their sadness and their power and will to survive. I've known poverty and cold and deep, gnawing hunger, and the impossibly dark well of loneliness, and a belief drummed so far into me since childhood that the word 'worthless' is scratched into my bones. When you have got nothing to lose, you have nothing to lose. Having nothing to lose is a real thing. I had nothing to lose but my anger. I have been a drunk, I have tried to kill myself, threw myself off a cliff edge, drag-raced motorcycles in the dark and under the influence, swallowed pills and alcohol to surf the edge of death to see what it was like, and waited to see if it would take me or if the dawn would come again. Then I lost my fear and I lost my anger, too, and became nothing. And the answer – the only answer there is — came and, as much as I tried to regain my anger, the answer came again and again: 'Be kind'; be kind to yourself, be kind to others. For the rest, there is no knowing, and where there is no knowing there is no point asking.

Camus, who stands in this absurd world and points out its ephemeral character . . . who 'seeks his way amidst these ruins', asks in *The Myth of Sisyphus* why should we live life and not just end it all? Towards the end of his ramblings he compares the absurdity of life with the story of Sisyphus, who was condemned to repeat the same useless and meaningless job of pushing a boulder up a mountain, only to see it roll back down, and was tasked to roll it up over and again for all eternity. Camus' conclusion? 'The struggle itself . . . is enough to fill a man's heart. One must imagine Sisyphus happy'. And I can be a flower in the garden, knowing that I will fade, allowing myself to blossom and fade, and loving that process. I found my God, and my empty-headed love, within the petals.

Anemone to *Zantedeschia*

All I desire is to make it more beautiful. I'm looking at seed catalogues and writing a shopping list in my notebook. I like an informal garden, unlike Proust's gardener, who frustrated Madame Proust with his insistence on straight paths. There is something Saxon or Roman about straight lines; something to do with dominance, control, ego and man's fantasy of mastery over his environment. Machines produce things of straight lines. We try to organise things to our liking, but it never lasts. We try to understand the things that we see and explain them with our limited words that fly by, like clouds as the sun rises in the vastness, and we fail dismally. I wonder why we bother. I am a Celt, and Celts do not do things in straight lines. When we go from A to B, it is not the A or the B that matters; it is the leaving, the going and the arriving. We meander like the knotty, twisted branches of the mountain ash, like waves and eddies and bodies and bones. Adapted to fit and work with all of nature's things that impact on our flesh and thoughts.

Curves are full of interest and surprise. Perhaps in places this garden is a little too chaotic, to a structured way of mind, and when Miss Cashmere comments that a larkspur is too tall in front of the corn poppies, or some such, I'll pull it out and throw it on the compost heap to make her happy. But I enjoy the chaos and the mixed colours and the insects they attract, so I sow purple larkspur and red poppy and honesty in the little border by the wall, next to the steps that go down to the path.

I'm planting the red tulip bulbs that I bought for Sylvia Plath among them – they will come before the annuals, and then die down as the annuals begin to flower. That's the theory; it all depends on the weather. Last year's annuals are dry stalks and seed heads, and I leave them to set their seed and gather the frost in their folds and cracks. I'll cut them down in the spring. I want the bed packed. It is a crazy little bed – my kind of absurd – and I can see, by her expression, that Miss Cashmere thinks it is a 'mess'. But I keep on doing it. I think that she will come to appreciate it and the butterflies it attracts, the flying flowers. I like to make her happy. In *Swann's Way* Proust says, 'Let us be grateful to people who make us happy, they are the charming gardeners who make our souls blossom.' I am her gardener. These plants that die in the winter, that I have to replace annually, are my stone, this garden is my hill; I am Sisyphus, and I am constant and happy. This cycle is a blessing, and the constant act of polishing the garden is a meditation, like a monk cleaning the Zen rock garden; daily picking decaying leaves from between the tiny stones, then raking the stones into waves as he returns the garden to its perpetual state of serenity, helps him to achieve his own state of serenity. Sisyphus serene.

There's an old fallen tree in the woodland at the bottom of the garden before the fields begin. It is full of the beginning of nature. Fungus, bugs, decay. Completing and restarting the cycle of life. Silent grey drizzle follows the morning's red sky. Everything changes. The seed has been cast, the dark heavy months will pass and new life will begin. The geese have flown, a dew-soaked web sags between two dry poppy stems, a beetle trots by – a moment of thoughtless calm, everything as it should be; a sun-bleached spiral of empty shell by my foot, a rotted-out stump that creates a pool of water, a nest that the ivy shades, larvae wriggle looking for the light.

The nights are drawing in. I am digging in the half-light, splitting calla lilies (*Zantedeschia*). Rose leaves cleared away, and the spring bulbs planted deep. Anemones sway. The brief shower passes by. I am wet, I am smiling, I shelter in my little white and shiny van, eating boiled eggs.

The Great Riddle of the Self

It is Sunday and I woke early again. We have domestic chores to do. I am not good with domesticity, I never wanted a home or a house to look after. I would be happier with a caravan, or a shed in the woods, or best of all a succession of hotel rooms, a passport and a wad of cash. I have the blood of travelling people pumping through my own internal highways.

I see the sun rise each day and, whether I am lying in our bed or sitting alone and watching it from the dark kitchen, it seems as if a goddess has come to me out of the darkness and washed me clean of the grubbiness of sleep and offered me the world beyond the glass. But today she is hidden, wrapped in seductive layers that glow inky blue-grey, and she seems to say, 'Stay, stay indoors for a while and just watch me. I am not in the mood for playing with you today, I have other things to do, so stay and watch and adore me.' It is cold and dark and I stay where I am, with my hand on Peggy's hip.

Forty-five years ago when I read Richard Hittleman's *Guide to Yoga Meditation* I learned to sit and look at a candle, then close my eyes and try to hold the image of the candle on the inside of my eyelids. I practised this for months, years even. That white flame is burned into my

347

retina and my memory and now, nearly half a century later, to stop myself turning and being restless and waking Peggy, I look for the white flame on the inside of my eyelids and, as it slowly materialises, I gaze into it; and as it slides off, usually to the left, I bring it back again and keep gazing into it. I drift off to sleep. In sleep there is no 'I', no separate thing called 'I'.

I still have that book, and an hour or two after I am pushed out of a second sleep's sticky warmth and the 'I' is born again, I wrap a towel around myself and go to find it on my shelves. It is the 1974 eighth edition, which I bought new from Grassroots bookshop. I abuse books because I love them, they are tools that are shaped by how I use them; the pages are falling out and yellowed, some passages underlined, others with the corners folded. I go to the first folded page, Chapter One, which has the title 'The Great Riddle of the Self'. A passage that I underlined all those years ago reads: 'We tend to believe that what we see about us is not only real but permanent. This is not true.' This sentence was the beginning of my journey and now, after years of practice, I can cut the ties that bind me to this world of things and float; hover softly above my body while chaos flows by.

I don't have much actual stuff left from my past, but that book and its simple teachings are one of the handful of things that I still have: a paperback beginner's guide, with old grainy black-and-white photographs of a muscular young man and a pretty young woman with a ponytail in leotards, doing Up Dog and Cobra poses. I haven't read it for decades. I don't need it any more, but it is kind of talismanic; it reminds me of the important things.

Reading Patti Smith's *M Train* prompted me to go back and read *Our Lady of the Flowers* again. I'm revisiting books I read long ago. Does that mean I'm now stuck in time, like other old people – back in the 1970s, where my reading life began – going round in circles instead of moving forward? And so I pull out a Tarot card from the pack bought way back then. The random card is the Five of Cups; it tells me that I am looking backwards, mourning things lost, rather than enjoying the prizes that I have.

No matter, I decide to enjoy reading Genet again, this time from the viewpoint of a much older man. I'm a different person now, so the story will be different. Online I order a new copy to collect from my local Waterstones bookshop. The book pile next to my desk wobbles a little, so I split it into two piles of more or less equal height and put my coffee cup on one of them: Jorge Luis Borges's *Labyrinths*. On top of the other pile is Jack Kerouac's *The Dharma Bums*.

The Dharma Bums reminds me of the times I have spent alone in the local mountains – my familiar Brecon Beacons and, further afield, the moors of the North, where I wandered. Missing that sad, lovely lonely feeling of freedom, of not seeing or hearing another human, being able to see from my vantage point that there is nobody for miles around. Being able to see into the future as the weather moves towards me. Feeling my connection with the landscape, without the distractions of human chatter or verbal clutter. I close my eyes and find the candle flame, then look for the grassy saddle of earth between two hills where a few hardy mountain sheep graze and watch; warm myself in the sun between the peaks, listen to the clear, icy water falling through the rocks, feel the crisp mountain air on my skin and in my lungs, then open my eyes, knowing that I do not have to be in the wilderness, because I have the wilderness in me whenever I need it.

Haiku

I am planting tulips in a bed that is full of the seed of summer's forget-me-nots (*Myosotis sylvatica*) under the light shade of a crab-apple tree. I am picturing the deep-pink tulip heads floating above a sea of tiny blue flowers and hoping they will both flower at the same time. It is a little late to be planting tulips, but they will be fine; they need a period of cold dormancy in the ground, and they will sit there resting over the winter and, as the soil warms in the late spring, as the daffodils are flowering, the tulips will be pushing their hard, red-beaked leaf tips through the soil, ready to take over after the daffodil flowers have gone. I write a haiku about it on my phone and tweet it:

> here in the present
> distracted by the future
> when I plant spring bulbs.

Planting done, I turn to the leaves that are falling from the acers that grow in the lawn. Their golden glow can be seen from the house and almost lights the sky above them. These really are the golden days. The most wonderful of days, when life is churning its vast mill.

If the leaves lie on the grass, it will be deprived of light and the grass will die in patches, so although the colourful leaves glow against the grass, I need to let the light and air get to it. I get my big wooden hay-rake from the van and start to move the leaves into piles. It was too windy last week, but now it is still, cold and frosty and still. The haiku keep coming, as they do once they have started; it is a meditation that stops the whirl of chatter.

> under bright acers
> in the cool damp autumn shade
> crops of dark toadstools

> whatever you build
> nature burns it down again
> beginning is end.

Breathing in and out, an unbreakable rhythm, I am raking leaves in filtered golden-copper light as nature burns itself down again. I have done this same task in the same way, under this same tree, every autumn for as long as I can remember. Nothing has changed – we have got older together. We play with each other and, ear protectors on, I start up the leaf-blower. Smiling. Leaves fly here and there as I chase them like a puppy with the breeze. Blowing them into rows, raking them into piles. A cold day of bright trees. Moving the leaves makes me warm. In a Zen frame of mind, I leave a few leaves scattered on the ground; there is no place for man's endless search for the illusion of perfection here.

I am hungry and the job is done. I have created a balance between the garden's needs and my need for beauty and harmony. I go to my van to eat – an apple and some strong Cheddar eaten off my pocket knife. A meal that I have eaten outdoors since I was a boy.

Gipsies

A homeless man is sleeping in a doorway in my village; he has been there for a few weeks now, his name is Gary. Last week I took him some things – a blanket, a rucksack I don't use any more, a hat – and I asked him how his night had been, and he said he had been fine until about 2 a.m. when it started to rain, and after that he sheltered in doorways. He told me how, a few weeks ago, he had been sleeping in a tent down in the scrubland, by the river, behind the cathedral. He came back one day and somebody had set fire to his tent, the photographs of his children, his spare clothes, his sleeping bag; his few possessions were gone, burned, shrunk to nothing. That is how it goes – when you have nothing, the few things that you do have increase in personal value, but disappear very quickly. To others, they have no value, they are rubbish.

He would have been sleeping under the bright-pink and showy white viburnum blossom that grows all around there; the thick ivy that scrambles over the stone walls, with their massed mace-like seed heads right now – groups of green bobbles on stalks. There is a winter-flowering cherry there that all summer long is unnoticeable, then as soon as the leaves are gone in the autumn, it surprises me and bursts into hundreds of pink-and-white flowers on bare twisted stems.

I remember John Clare who, along with the Welsh poet Edward Thomas, does not live on my bookshelves, but on my desk by my right hand. Clare wrote in the late 1700s the most perfect poems of fields and birds' nests and 'of men disdaining bounds of place and time' who 'plod upon the earth as dull and void'. He was an agricultural labourer who wandered and found on his return home that 'everything seemed strange to him and altered'. The book Miss Cashmere gave me told me how he spent the nights with gipsies, whom he wanted to join, but he woke one morning to find that they had moved on quietly in the night. Sent to an asylum, he walked back to his home, taking three, three and a half or four days, depending on which account you read, thinking he was Lord Byron and eating grass. He was to walk eighty miles, intent on seeing his wife. She had been dead for three years. He, like many poets of his kind, was mouse-poor but well read. He had some brief financial success as a poet, but it was not to last. Subject to bouts of depression, he ended his days in the madhouse; and I wonder at the resilience of humanity that more of us do not end up insane, or perhaps we just don't express it, keeping it all inside us and living lives of quiet desperation.

Last night there was some snow – not much, a half an inch perhaps. A thin, cold drizzle is falling on it, making tiny holes. Underneath the snow is melting already. A pair of blackbirds are chasing each other through the rosehips.

They seem happy. They have no shadows. I give them half of my breakfast apple and I stay home to read. I do not get dressed today, but wrap myself in a rough Welsh blanket with a pattern of bright abstract stars on a scratchy black background. Another personal relic. I sit on my sofa by the kitchen window, read poetry – the language of the birds. I make soup.

The Lily Gardens

The sun rises late and it goes dark at 4 p.m. The days are short and evenings long, and I wonder what to do with myself after eating and before sleeping. It gets dark and I am not tired, as I haven't used my body enough.

The snow has melted. The sparrows are singing, even though it is 7 a.m. and not yet daylight. The sun stays low and takes its time, barely lifting above the grey horizon. In the garden the work is ending, and I am stretching it out to fill the days and weeks. It is too early, too dark to work, so I sit in my kitchen, drink hot chocolate, listen to the little flock of sparrows that flit between my own tiny garden and those on either side. We all want to own them, we all feed them to lure them into our space, and they are the happiest, fattest sparrows in town.

I have at least two hours before the day is light enough to work in, so I take a look at my ageing, faded, broken copy of Sackville-West's *In Your Garden*. I open it to the pages for November and read about planting alpine strawberries. This reminds me that I need to pot up the suckers from the strawberries and bring them on as new plants.

This little book, published in 1951, has become a classic. It is a collection, amongst other things, of Sackville-West's gardening articles from the *Observer* newspaper between 1946 and 1951. At the end there is an appendix, three pages of long-gone nurserymen; just the names and addresses of nurseries where one might have bought the plants she recommended in 1951. There are no telephone numbers, and I suppose in those days people would write a letter to the nurseryman. I pick one at random:

W. A. Constable,
The Lily Gardens,
Southborough,
Tunbridge Wells,
(Lilies a speciality).

I search the Internet to see if he is still there. I can find no trace of the garden. An online map shows me that Southborough is now a modern housing estate. But I do find a book, *The Modern Flower Garden: Vol. 6 Lilies – With Chapters on Lily Species and Propagation* by W. A. Constable, described as 'containing classic material from the 1900s' and published in 2010 and again in 2016 – a reprint of an out-of-copyright booklet; and there it is again at an online bookseller as both an e-book and a paperback. Mr or Ms Constable has been resurrected in print.

My flower-reading is taking me into exotic places; *In Your Garden* has led me to read Sackville-West's nature poetry and learn about how this seemingly correct young lady had multiple love affairs, perhaps most notably with Virginia Woolf, for whom she was 'Orlando' – who, like Tiresias, changed gender halfway through the story. Flowers and a lusty nature seem to go hand-in-hand.

My youthful, burning fury when I discovered these celebrated people, with lives of privilege, play and freedom from toil or hunger, faded. I am – was – a poorly educated working-class man who left school at fifteen, a one-time destitute. I was taught by my peers and 'betters' to accept inequality and the class system, and then I learned to be angry about it; and then learned that anger is self-harming and that we are all merely expressions of a creative universe that has no pity, does not notice anger and asks for no forgiveness. Now, I simply think about the garden and the flowers in it, and the day as it passes. Those privileged people are flowers in my garden just as much as Jean Genet who, degraded and destitute as I once was, is a flower. Genet, too, regards his co-inmates as flowers, the flowers as his family. Genet, a beggar, prostitute and thief who stole from his foster carers, rebelled against their values and tries to subvert those of his readers. He was imprisoned from a young age for petty theft and lewd behaviour, and had smuggled out of his prison cell the most passionate and visceral, lust-filled words as he waited to be transported to Devil's Island (although he never got there). Genet glories in how debased, how petty and dirty, he and his lovers are and raises their degradation to the highest ideal – a state to be adored, worshipped, aspired to – as he writes masturbatory fantasies about his co-inmates in *Our Lady of the Flowers*. He delights in wrongdoing, and I delight to have him in my garden. Life for humans is as fluid as it is for a caterpillar, if you allow it to be. He became close friends with Cocteau, Camus and

Sartre, who wrote extensively about his works, as he became a leading figure in avant-garde theatre.

The leaves are curling on the ground, with frost crystals growing on their edges and along the raised veins. I suddenly have a desire to take in as much as possible, to collect the colours.

I've planted up fifty strawberry suckers, and my hands smell of strawberry. The leaves, hairy and slightly sticky, changing from green to red, turn my hands green. The old plants ripped out of the bed are thrown on the compost heap to die and rot, the younger plants dug up and replanted in rows. I don't like rows, and they won't stay in rows for long as they grow and spread, but somehow rows seem the right thing to do with strawberries. I clear the few weeds from the raised bed and till the soil, and top the bed up with compost from the heap until it looks tidy. I am feeling tired as the cold creeps in, and consider packing up for the day. And then, thinking about the long, empty evening, I decide to do another hour until I can't see properly. These are the dark, bleakest months as we wait for the equinox around 22 December and there is not much to do, but I usually manage to find some little task.

I snip off some of the mahonia flowers. These spires of small yellow flowers above hard and sharp holly-like leaves on stringy, tropical-looking trunks flower from November to February here, and on a mild winter day will be visited by bumble bees. I take enough for a small vase and tie the stems together with the musty green garden twine that I carry in my pocket, and leave the posy on the wrought-iron table by the back door. A room without flowers is a room without lust, and these spikes and soft flowers look good in a vase and scent the room with a lovely lily-of-the-valley-like scent, although they don't last very long.

As the light has gone, I go back to my van and drive home, stopping off at the supermarket to buy booze made of wormwood, juniper, gentian, fruits, seeds and roots, stems and leaves fermented with yeast – the oldest culti-vated life form known to man – and barley seeds allowed to sprout on vast damp floors, then smoked over burning peat, itself rotted ancient plants; distilled with mountain spring water and tended and blended to create celebratory delights that intoxicate and bring out playfulness. I head home to Peggy, with a bottle of gin for her and a bottle of whisky for me, to shower and change, mix drinks and start my own small blaze.

Lifting Dahlias

I kneel, a gardener priest tending to my flock of dahlias, gathering them in. A reaping frost bit into them deep and hard in the night and took their colour and backbone. The water in their cells froze, expanded, burst open their walls, their structure gone; and as the sun rose and the ice in their veins melted, the plants lay down, blackened and dead. Yesterday's massed summer flowers are slime already, steaming under the clear bright sky. All along the border, one after the other, I snip off the broken top growth with my secateurs and then, with a border fork, dig into the frozen earth and gently lever out a lump of soil and root, leaving them on the top, letting the sun soften the icy soil so that I can clean the fat old tubers and store them away in boxes of sand for the winter, to be planted again next spring after the threat of frost has gone. It takes half a day to dig them up. There are slugs hiding in some of them; wet and nude, they lay clusters of pearly eggs. I leave them on the top for the birds.

For the plantsman, frost is a problem. For me, it is the earth looking after itself, the soil sucking back in the nutrients it has breathed out, breaking down the fallen leaves and turning them into earth; water expanding and contracting in between the earth's grains and cracked rocks, breaking up the surface of inanimate, porous things, killing off the bugs so that the young can hatch, fresh and

362

new, stopping disease in its tracks. It gives me an opportunity to rest and wear wool, to build fires, to look at other things, to travel away from the garden that consumes my attention, and to breathe in again.

Leaving

The dahlias on the bed are drying out – it has been a sunny day. Small birds have eaten most of the slug eggs, and I brush the remaining tiny, moist pearls and soil off the tubers gently and lay them in sand-filled boxes under the benches in the dark shed. The brazen fuchsia flowers with flipped-up skirts hang from naked branches, purple and pink and red. A burst of pink nerines, in clumps by the wall, are almost frozen in ice. A few lonely roses flower, soggy among the masses of hips glowing red; a squirrel runs along the wall top, pulls a berry off the branches with his paws, tastes it, then throws it away before trying another.

Before the winter comes I am tidying up. Spring is the time for pruning most things, but I am cutting back some of the roses and buddleia. I will prune them properly in the early spring, but winter winds can rock and twist and break these woody plants and so I haul down their sails a little. Kneeling on the wet grass, I'm cutting a thick branch at the bottom of a buddleia with a Japanese saw, whose teeth point backwards, so that it cuts when I pull. There are new shoots budding at the base. Buddleia can be cut down to a stump and will still grow back. A lot of people think of it as a weed, but I like it. Buddleia do not care where they grow – in a wall, on a chimney stack; they are unconcerned with our feelings; they are robust criminals, and the flying insects that we all depend on love the massive drooping purple flowers.

I am warm in a moleskin shirt, a knitted jumper, my old tweed jacket. I stop cutting briefly to rest, as I notice a late leaf fall from a chestnut tree nearby, and for a moment watch a couple of them tumble in the still air through the mist. I wait for more, but no more come. The bright, flaming berberis leaves and the cotoneaster berries are shining in the distance; greyed out, as if seen through light silk, an underexposed photograph, the cloud rests on the earth. I look up the lawn towards the house through the gauze of winter mist, and in the distance I see the vague shape of a figure coming towards me. It is not Miss Cashmere, whose shape I would recognise anywhere, but it's too vague even to see if it is adult or child; just a moving forked form, swaying as it moves its weight from one leg to the other, rippling in the mist, large and growing larger – an adult. The ground is damp and the fog feels good to work in, cooling and fresh; like everything in life, it comes and flows around in its own individual perfect way.

Sawing is hard work and the branch lies on the ground next to where I kneel. Getting up off the ground is difficult for me these days, my knees ruined; so I stay down with the branch and the saw for a while, using secateurs to clean up the base of the little bush, moving fallen leaves off the perennial plants that grow around it, the lilies and hellebores. The figure approaches and breaks through the curtain. A man I have not seen before in this garden – my garden, Miss Cashmere's garden – and I feel totally invaded, threatened even. The man introduces himself as Dorothy Cashmere's nephew. As I kneel before him, he tells me that he is sorry to say that Miss Cashmere died three weeks ago. In hospital. In October.

'I am so sorry,' I say to this man I have never seen before. I don't get up off the ground.

He says casually, to my back, that he would like me to keep 'doing the garden' – as if it were the laundry or something – 'until the house is sold'; the 'new owners might want to keep you on,' he suggests, as if he thought this might be my most pressing thought.

'Thank you,' I say as he turns and walks away.

'I will pay you as normal,' he says, his back moving away from my kneeling one.

I skirt around my feelings, unable to think deeper; perhaps I might unearth a hurt that is locked safe away from language. There is wonder. And I begin to think these words, and, oddly, to think in words:

> . . . I have been looking for her for weeks
> Then
> . . . I won't see her again
> Then
> . . . I was looking out for her each day but she has been burned already
> Then
> . . . or buried
> . . . but will not rise again

. . . like a perennial.

Then

. . . I've been looking for her for weeks but she has been gone

Then

. . . She has been gone.

This empty monologue harasses me, as it cycles and re-cycles. I crash into my mindless work and nothing is done well, so I shut the words down and try to rise above them; they can only say something of sadness, or increase the sadness and repeat themselves meaninglessly. A heaviness invades me, populates me, and I not only allow the word-less, leaden melancholy to grow, but actively embrace it as a comfort. This weight of feeling seems impossible to do anything with but hold. Could I be without it? I don't want to be without. Where there is a feeling of love, there are always feelings of loss; and I may feel both or neither, but cannot choose one without the other, and I realise that I love her.

My warm breath makes clouds. I look up at the clouds. The crows high above, circling. From my kneeling position I decide not to be earthbound; I bend my head back and crack my neck and look up. There are no warm, fluffy blankets on this flower bed where I kneel, where flowers sleep and I do not. On my knees, I am neither praying nor begging. Without thinking, I uncurl, stand and extend, bend and pick up my tools: the trowel and hand-fork, the wooden handles worn and splitting and shaped to my hand; the steel blade and tines pitted and sharpened and polished by the grit in the soil, one end in me, the other end in the earth. I put them in the old mud-coloured canvas bag on my shoulder, resting on my hip and swinging a little, as I move my leg and limp, empty, to my van.

December

We barely spoke, I tell myself . . .

I stay away from the garden for days. She was just an employer. That's all, nothing more. We barely spoke, I tell myself several times. Yet I grieve as if she were an old lover. I tell myself that I'm sad about losing a way of life, a routine, but it is something more. I have lost a world and there is now a hollowness that is hard to understand.

Death is a once-in-a-lifetime event for an individual – a momentous one that we try to avoid; and yet, as somebody who has lived with it, I have come to trust it as a friend and adviser and have no fear of it. I have come very close to my own, and have watched life leave living bodies at my feet, in my hand, in my arms; seen the open eyes look at me, then close as the neck goes limp and the body gains weight when the muscles sag, the lips, beak or snout relax. Animals, birds, people. It is nothing to fear, but even the death of a small mammal leaves me with an empty space that, while they lived, they created, warmed and filled.

Whether you are a wallflower, a lover, a drinker, a dandy or dancer at this party, as the night comes on, you fade away and leave. As I think about drinking my fill from this cool stream that will not run for ever, the rain starts; a few drops run down the window, then the sky blackens and all detail is gone, and the rain is gravel flung on the glass and the wind is furniture being dragged across the planks. The house feels like a ship lost at sea in a gale, in full sail, pulling us along to the end of the year, to the end of the Earth. Even though the windows are cracked open only half an inch for ventilation, the blinds are rattling, the wind is playing with them like a cat and knocking things off the window ledge.

Inside, with hot chocolate and book, and Peggy next to me, the world seems rich and more full than I can stand. Full of the emptiness of love / loss, full of weather. I put down my book, close my eyes and take in the sounds of sea and gale-force winds, and the smell of wet air coming in through the cracked open window, while the black sky turns to white and silhouettes the bending, madly waving trees. And suddenly the wind rests, stops spinning, and the poplars in the distance gently sway this way, then that, as the storm passes and the rain softens and settles into a steady patter that will last the whole morning.

Later Peggy and I are in our kitchen, doing kitchen things together. Making steam. All the seed that will be sown has been cast. The geese and flowers and swifts and bees have gone, and the dark and heavy months are here. I still hold my love. I still hold my love. I still hold my love. It is Saturday and we will spend the day together. Far away in the town a moving siren on a vehicle sings its wailing song, which always means change; it sings of hurt or fire or the crimes of some poor soul who runs, for stealing from some poor other soul. In the past the singing sirens on the wave-lashed rocks, monstrous birds with women's heads, seduced the passing sailors; men were so easily entranced by their beautiful voices that they dived into the sea to reach them, only to die among the piles of rotting corpses of previous victims. In her poem 'Siren Song', Margaret Atwood writes it as a different cry, one that a man might find impossible to resist: the cry of a woman asking for help, as if the listener was unique, the only one who ever could help. It works every time, she says.

In the cracks between those fearful wailing sounds, I hear a blackbird warn a prowling cat; a robin sings into the vacuum, and a thrush; a helicopter closes up the chasm, to hum and suck and drown the siren song. At the same times every day, at the school just behind the houses, the kids chirp out the hours to everyone around, like cuckoo clocks as they come out to play at half-past ten and twelve and half-past two. At three-forty a shopping trolley rumbles by; a child in red school tights is softly speaking *Cymraeg* to her mam. Then, on Sundays, quiet; no cars, no birds, no children, another chasm. And so the days roll to their end, with my own internal concert: constant singing, heartbeat, breath and squealing tinnitus and love.

There is a Christmas cake in the oven, and Peg and I will celebrate the winter and the beginning of the renewal in our kitchen, thanking what has gone, washing our hands of it and welcoming what is new. Celebrating our children and their anarchy; grateful for contraceptives and whisky and wine and cheese and wood for burning and fresh air;

the sound of rain and wind through leaves. Lifting my heart in a glass of water from my tap, transient, enduring, old and cold of iron and earth and ancient rock. Some drunk, some spilled and mopped away. Remembering the peat-stained water that flows from the taps in Kintyre and wishing I had a little jug – a few drops for my whisky. The postman brought me poems from the Argentine, dried sea vegetables, a small cheque and a bill for a similar amount, maps and other number-poems. A peanut fell out of my beard and now I can't find it. I'm distracted by the curly hair on the nape of her neck.

Back to Work

There is thick winter morning mist. I kneel to weed around the raspberries and it hurts to get back up again. My knees hurt and I have to crawl to the end of the row to pull myself up using a spade, embarrassed but pleased there is nobody there to see. I am too old, this constant labour is too much. The wintersweet that I pruned in the spring, and which sat in the background all summer, is showing tight buds that will flower over the winter. I will not see it, she will not see it. Nobody will see it. Now it is nobody's garden. Yet I am here and for a little while, and for the very first time, it feels like my own garden; it exists in the hinterland between owners. The frost is cleaning the old and cankered, and a blanket of snow may come to warm the earth. For the first time in many years I think of my mother, who died when I was a boy and left me to fend for myself. And I think about all the older women in whose unwilling faces I looked for her as I grew up. And I wonder about my love that lacked desire for Miss Cashmere and perhaps, in her distant manner, I had found and lost again what I had been looking for.

When a song is ended, it leaves nothing but a feeling in those who heard it until that feeling, slowly moving backwards in time, collapses under pressure from more recent feelings and is replaced. Events become memories, unreliable stories, fade away at the ends. Unconnected and distinct from the day's experience, they become one of the millions of strata that make us who we are. We are the sum of all our experiences. We are waves on the ocean, interacting with and affected by all the other waves that move and die and are washed up on the shore. We are each a breath, a song, a flower. We are time itself, and mine has been long and I've collected many disconnected layers.

The bees have left the ivy now. The cold has sent them underground; the ivy flower balls are stiff and set to drop their seed. My shadow bent, I rake the last leaves from the lawn beneath sleeping bare branches and look to the sky for more snow. A blackbird's small shadow, flat and black against the earth. Shadow, earth, bird, black. Fixed like a hard photograph. Only the beak-gripped worm moves. I crush a handful of the bright leaves from the Mexican orange blossom (*Choisya ternata*) as I pass and sniff the powerful scent on my fingers. The powdery, thorny arches of white-stemmed bramble glow next to it as the low sun hits. The bright-red dogwood, whose branches I cut to take home for Christmas decoration, is bare of leaves and looking fine and it, too, glows. This shrubby area is not my favourite, although the plants here shine in the winter sun.

In the afternoon's thick fog I trim the frosted beech hedge, making it flat on the top and sides again after a summer and autumn of growth. Eight feet tall, six feet deep and the mist is so thick that I cannot see either end. A few crisp brown leaves curled and matt and hanging on; behind them hard, shiny brown pointed buds, needles, tiny cigars of tightly rolled and packed new leaves, ready to open and stretch out when the warm sun shines on them, always much later than I expect. Quenelles, spears, cocoons.

Dozens – maybe tens of dozens – of gardeners, long gone, dead and moved on have cut this old hedge. It is lonely and cold, a lone bed and blanket. Tucked up with my back to the world and my face to the leaves, can I feel their presence? Those who have passed? A shadow of a memory behind me, perhaps. Nothing more. I can see old saw-cuts deep inside the hedge, made by gardeners long before me when the hedge was only two feet deep and even then perhaps twenty years old. I only believe in what I can see and hear and taste and smell and touch and know, but nevertheless I become aware in this tight grey hidden nearness that they existed and are gone, and now I exist and will be gone, and then another will exist. Kneeling by the flower beds, a battle-scarred, staring, starry-eyed gardener. Looking out to avoid looking in, and cutting down the dead and decaying to keep it tidy.

A poppy head, its sides crumbling, its seeds long gone into the breeze, and I want to freeze it; it is so perfect I want to hold it at this stage of decay for ever. But I cannot, for it needs to continue and be gone in its own time without my interference.

At the bottom of the garden there are trees I planted, oh so many years ago: hornbeam and cherry, crab apple, marginal trees that live at the edges, and beech, oak and silver birch, copper beech that live in the middle. Some of the trees have fallen in storms already and been converted into logs, leaving small clearings. Others never grew beyond the two- and three-year-old whips that I planted; others have thrived; and holly and ash and willow have come in, brought by winds and birds, and into the clearings brambles and nettles and ivy have come, which provide food and cover for wild birds and ground-dwelling mammals.

Every year I cut down the brambles to prevent them taking over, but otherwise I have fondness for the way they grow beneath the trees where birds perch and in their droppings from the fruit they've gorged on sow the bramble seeds. They grow fast, with backward-curving thorns they clamber over grasses and wild carrot and cow parsley and thistles; and when they are long and heavy, they arch over until the tips touch the ground and grow roots, and this new plant sends up three or four shoots that spread out from the centre and begin their reach and fall. And so the whole area becomes a cathedral of arches, which in the autumn is decorated with blackberries. The stem is hexagonal to keep it stiff and strong, six flat sides and ridges along which the thorns grow, sometimes in pairs, sometimes alone. The sunlit side of the stem is purple, the shady side green.

Underneath these thorny arches, under the green thorny roof, there are bare shaded spaces where nothing much will grow but a few straggly grasses, where foxes lurk and rabbits and hedgehogs and wood mice shuffle and run, where ground-birds nest and feed on the blackberries. I decide to leave it; I will not cut it down, will let the animals have their church. I have the chance to leave this wild place alone, to stay out of it. Perhaps I'll never go to the wild places again; perhaps these places are not for us. There are flowers everywhere, in the cold soil; winter-flowering cherry trees covered in small pink blossoms, witch hazel with yellow-brown flowers that look like dead spiders lying on their backs with stiff legs in the air, hellebores with risky pink flowers, daffodils poking their tips from frozen soil, little Welsh-poppy seedlings, pretty tiny seedlings. In this ageing garden, where there is always something sprouting, there is such a thing as perpetual youth.

It is the end of days, for houses and gardens like this, and I'm mindful that I am approaching my own end with a rapidity that only becomes noticeable as I catch a glimpse of the finishing line. It has been a long run and I am tired. I was told that I spent the first two years of my life crying. I didn't want to come into this world, and now I don't want to leave. 'Laugh and the world laughs with you,' my nana used to say, 'cry and you wet your face.'

It is cold, and now the sun that glowed through this flimsy fabric of a day falls behind the trees, the thin cloud in the icy sky turns opal and jackdaws circle in a great flock – the young now confidently part of the tribe and fending for themselves. And I am just an audience, a witness.

The Floating World

I turn up at the usual time; it isn't light yet. A writhing in my chest as I drive to work. I don't want to go to work, I never want to go to work again. I wander around, note a few jobs that need doing, not many: a beech tree with a fallen branch, the brambles. The pile of slowly curling books, still fading, falling to pieces in the greenhouse. Nothing is ever finished. I decide that I have had enough. I want to emerge from this piece of earth like a caterpillar, rise into a tree, grow wings and fly off. I want to be reborn, to rest and sit in pavement cafés in a dark suit, dress like Philip Larkin, go to see Paris, Rome perhaps, the south of France. I want to wear a bow tie and write poetry. I write a note on a page of my notebook in my childish block capitals: 'I AM SORRY, MY HEART IS NOT IN IT ANY MORE.' I tear it out, wrap it around the key and post it through Miss Cashmere's letter box as I leave this world and its endless cycle of birth, life, suffering, death and rebirth that has been my lightness and my eternal rolling stone.

I clang the gates shut, click the worn old padlock into the chain and leave without my key. I am crying, actually sobbing. I've watched the trees that I planted grow enough to shelter me from the sun while I ate their fruit. Grew roses from cuttings that I cut from roses that I grew from cuttings and watched them all bloom, sprayed them and pruned them, and tied their flowers with my smelly green twine into bunches for Peggy and Miss Cashmere. I made a meadow grow from a swamp, to blush and flower each year as it grew more mixed and varied, and brought in insects and mammals and birds that were never there before. How many years? How many? The number doesn't matter; numbers mean nothing. Fifteen years doing something you hate is far less important than two years doing something you love. All of time is short, the depth of time is bottomless. I will never really leave this place – it's part of who I am, part of my dream. But I cannot do it any more, my body cannot do it any more.

I hope nobody enters this floating world again, and that this dream becomes a real place with ash trees growing through the broken greenhouses; the solitary beech tree in the lawn becomes a forest as the seedlings it makes remain; unmowed, the open lawns become meadow, the flower-beds merge and spread, and wild roses meander across the tops of hedges, burying them in flowers and hips and blackbirds and thrushes; the pond breaks and floods and makes a water meadow where frogs and newts hide from dogs and cats gone wild. When the last person leaves – the very last, leaving a new Eden, when the power goes off – this world will still be here.

Home

In our thirty-five-year-old marriage bed, we see the curtains flap in the breeze and sunlight spikes the wall. There are sunbeams. We read together and wait for the cat to come and see us. This is home. Whenever I use the word I become a little uncertain of it. When each day I wake, lost, I turn to her and find my home. Whenever Peggy is next to me I am home, wherever that might be. I try to follow Rilke's advice and not write love-poetry, but I am no longer young and cannot help myself.

Watching the clock hands move, then the clouds move against clear blue sky, with hot drinks. We are breathing softly, yet inside I'm tightly drawn, contracted. I look into the vortex in my mug. My old cat arrives, seeks warmth, curls on my fat, wide lap and we both give comfort. Then at last the clouds leave. Trying to read, but unable to stop looking at the clear, empty cold sky. Cold, I am wrapped in tartan wool while Wales looks wintery. I am watching for snow.

Nietzsche, as part of his 'eternal recurrence' idea, suggests that if, in your coldest, loneliest moment, a demon were to creep in and tell you that you would have to live this whole life over again from beginning to end – innumerable times, again and again with nothing new in it, every pain and joy again and even the demon's visit-ation again – such a thing would soon become unbearable

382

and most of us would resist, apart from those fortunate few who have lived a happy and comfortable life, who would see the demon as a god and say 'yes, please'. This life is beautiful, but there is no such beauty in eternity; its glory is only because it ends. Looking back on those glorious years, in time they will collapse into a sweet little story.

Although I have not always lived a good life and it has rarely been an easy one, I have become happier as I have aged. Nietzsche goes on to say that one who is lucky enough to grow old and see two or three generations is like the man who sits in front of a conjuror for two or three shows and sees the same tricks over and again. Of course I have seen the young getting excited about the things that excited me, and creating things that have been created before, then settling to have children who repeat the story. The serpent eats its tail, the waves wash up on the beach, the tide comes in and out, and everything comes round again, as natural patterns reverberate through the generations.

I think before I die I want to see a few more tricks, some new ones. Just for the fun of it, for the greater love of it. I may have another twenty years of a different life altogether if I am careful, and perhaps for the next ten I may still be able-bodied.

For a few days, uncertain about what I should do next, I go for walks and watch television. The walking was good, but the television left me feeling hopeless and worthless. The vile psychopaths it wanted me to enjoy could not hold me in their spell. I couldn't bear to absorb those indelible images and pollute myself with hatred and violence. What didn't depress me made me feel like the cornered victim of a wealthy bore with their holiday movies, who sucked the vitality out of me. I pulled the plug and took to my bed to read, cocooned again in poetry and wondering about a rebirth. If I go to sleep a slug, do I have to wake up a slug or can I become a butterfly? Can I be a cloud that has no weight? Can I be a drifting strand of web that has no meaning? Can I be a falling sparrow feather that has no purpose? Can I be a constant hanging mote of dust caught in sunlight? Do I still need to be somebody's flower? There is a flow, and people who try to go against it are destined for disappointment, but I don't know where it goes. I could perhaps become again something else?

Does a chrysalis dream as it changes? I went to my books. Randomly I took out *The Collected Poems of Dylan Thomas*. I opened it by chance at page 150, where there is a four-line poem called 'The Train Journey'. It is short and sweet and reminds me of a poem by Günter Grass that I used to read to my children when they were little, a tiny jewel called 'Happiness' about a bus driver. I spend the day reading Dylan Thomas. I read 'The Roads', 'The Bus Ride', 'Glasgow', 'London'. My subconscious lights up. I will not 'go gentle into that good night', not just yet; although I am prepared to go ever so gently when it does happen, I plan for it to be a poem. I feel apprehensive, and I decipher my apprehension as a sign that the right thing to do is to go towards the fear, to change, to emerge somewhere else, to be reborn.

I miss my flowers and my insects and the floating world.

Flowers

I nearly always wear a suit now. My daughter some time ago told me that I dressed like a schoolboy and I was too old to dress like that. I asked her what I should wear and she said, 'You should wear a suit.' I said, 'What – all the time?' and she said, 'Yes.' And I asked her, 'What colour?' and she said, 'Blue.' So I do. I no longer have work clothes, just suits. I do not do my day job any more. The roughness of the wool of my suit reminds me of my blanket that I carried when I was homeless, and I feel secure. Peggy is wearing a ribbon, so to show her that I have noticed, I am wearing a tie.

She has gone to the library to do some research for a book she is writing, so once again I am in a coffee shop doing my own kind of research, reading a book, watching people. It is spitting a little and the windows are steaming up. A Sikh told me once that everyone was a flower in the Lord God's garden – all the individuals, the colours and races, tribes and religions; it was an idea that I fell in love with, and kept coming back to over the years, and eventually I chose to be a flower. I don't believe in any kind of God; if there is such a beast, he has horns and hooves and plays the pipes and doesn't live in the sky, for us to look up to and worship, but underground, and pushes all the wonderful things out of the soil for us to admire, pushes us out into the world, then takes us back again to join the earth. A creator that gives us passion and music and lust: that's my kind of deity, should I ever need one.

I have a seat by the window and I watch the different people passing by; their colours spread and glow, smeared and lifted by the wet window like a field of passing peonies, camellias and dahlias. There goes a handsome lady in an orange top, with two equally proud spaniels on yellow leads, crossing the road and it's good. An old man who looks like he has seen a lot of weather passes by in brown tweed and a blue shirt and takes an empty Gordon's gin bottle from a carrier bag, green, and slips it into a waste bin, black with a gold council crest painted on it. I'm pretending to read my book while I wait for Peggy, and I'm watching the Japanese tourists who light up the place like a hundred candles on a birthday cake, like magnolias in bloom. Taking photographs of trees and whatever else is hanging about that's foreign and weird to them. And there are people trying to be fit, and people trying to be families, and people trying to have fun or work out what those things mean.

I am drinking an espresso, watching market stalls pop up across the road like instant window boxes in narrow streets. Feeling slightly ill, perhaps because I'm drinking coffee. Perhaps because I stayed up late last night, singing along to an Irish band in a bar. 'Wild Rover' and 'Black Velvet Band', and drinking Jameson's. I remember working as a shunter in the coal yard at Stockport station when I was about nineteen or twenty-two, and lovely old Paddy Mullen, a man with hands the size of coal shovels – he could hold a big brown teapot without using the handle and had a head as big and as ugly as a brickie's hod. I used to go out drinking with him and singing in bars; he knew all the old songs that I grew up with, and he had a string of widows that he used to tell me about. I laughed with him at his failed attempts to make sure they didn't find out about each other – having to nip out of the back door of the pub or hide in the toilet. He was such a distinctive man, and Stockport such a small town, that every now and then he ended up in a blind panic and lost one or two of his girlfriends, but always found another. He had been married before and was determined never to make the same mistake again. He was a sentimental old fool, who would cry into his whiskey when we sang 'Freeborn Man' and made me cry, too. Poor old happy Patrick, who just a couple of years away from his gold-plated clock got run over by a train and killed. He wouldn't have wanted to die like that, with his boots on; he would have said that he wanted to die 'on the nest'.

'I had a bloody lovely night last night, lad,' he would say. 'Spent the night on the nest with one of me widders – you need to get yourself a widder, boy.' Although we were years apart, we were good mates. The closest thing to a friend I had. Almost a father. Not long after that I changed my job, 'improved my position', became a guard and was on my own again, crawling slowly up and down the country from coal yard to coal yard in the night, in a swinging wooden brake-van at the back of a train, reading by lamplight, my only companion in the night a driver a quarter of a mile away at the other end. The railway was full of lonely men.

At the market across the road, a man with a Mohican and a Specials T-shirt is selling records next to a tall woman with a bric-a-brac stall. I watch them chat and share a laugh through the steamed-up café window. She comes in for a cappuccino; she is sixty or seventy with long, plaited grey hair. She looks like Patti Smith, she is handsome and strong, slender, she has a long nose, and I fall in love with her instantly and decide to extract every drop of juice out of this day by doing nothing of any value whatsoever. There's a small-mouthed woman in the café, who refuses to smile at her husband's good humour; her classic bobbed hair made her pretty at first, but her controlled tight lips look mean. She is handicapped by her point of view; he by his ignorance of it. We are all handicapped by our point of view – our perspective on things that gets added to every day of our lives. I try to live without a perspective on things, without a point of view, but it is very difficult. Some times it is easier than others.

A sixty-year-old man, marvellous and proud, with a thick grey ponytail, leather jacket, a fist full of skull rings. A woman in her twenties in a smock dress with a double-doorknob hairdo, and a fairy tattoo on the back of her neck. A hurrying woman in a spotty coat, hair black as a raven's wing, as tar, as coal, as a cassock, as a politician's lying mouth, as the fiery hobs of hell, all tumbling down her small back. They are all passing, the tourists and people trying to be families and such; and some pigeons and a few crunchy leaves drift by, and I'm wondering: Is there more to life than this – my days spent watching the passing? There doesn't need to be. This is probably enough. And I choose to join them, to melt into them, reflected in the glass, my shiny bald head, my big white beard, round glasses, blue tie, coffee cup to my lips. We each have a space that fits us, all our darks and lights; and when we leave, the space lasts for a while, then slowly closes up, changing shape as it does, and we are mostly forgotten. I have forgotten most of my own life – why should somebody else remember it? I have found a purpose in life, and it is to bloom, wander around, have a bite to eat. Nothing of any value lasts for ever.

Later, to amuse myself, I picture the passers-by with imaginary horns and they look magnificent, those goats and stags and deer and bulls. And I wonder about their labyrinths and mine for a brief moment, then think their journeys only matter to them; and mine, when it matters, only matters to me. A tremolo of rain in a minor key passes by and breaks to a second-rate sun, the clouds dissipate to nothing, heat comes and dries the pavement, and life in the market starts to buzz with activity and there's laughter. I finish my now-cold coffee and go to buy flowers, because I want some near to me. Wet stone pavements smelling of the wet-stone-pavement smell. I sit on a park bench. In the park there are four men dressed from neck to ankle in high-visibility orange waterproofs. The machines strapped to their backs are orange, too. They are blowing leaves. It is windy. They work for the Parks Department. The leaves inevitably do not stay where they are put and return to the path. Nevertheless, the men continue to blow and to chase in a steady workmanlike way and, despite their work being ineffectual, they do not stop. They seem happy. Sisyphus. I picture them all with massive antlers, a herd of calm orange deer, snorting slowly through blown leaves.

In the shop next to the flower shop I buy a blue bow tie with white spots and replace my striped tie. It takes me a few goes to get it right. I haven't worn one since I was a posy art student. I am sitting on a bench by the library, holding a bunch of yellow roses and waiting for Peggy to come back; we are going to go and drink gin in a bar. A man comes up to me, he is a white man with dreadlocks and baggy striped trousers and a colourful shirt without a collar – hippy clothes. I think that he looks like a man from Bristol. His horns are like mismatched and twisted cow horns, and he looks at me sitting on the wet bench and says, 'What are you?' He is not aggressive, merely curious, inquisitive like a goat, so his horns become goat horns and his eyes become intelligent. The goat is my favourite animal. He wants to know why am I wearing a suit and a bow tie. He asks, 'Are you one of those "chaps" from London?'

'I am just a man waiting for his wife,' I say.

'Why the bow tie?' he asks.

I answer, 'Why the baggy trousers?'

'Because I like them,' he says.

'Me, too,' I say. (I don't mention his horns.)

*

A broken cloud of small birds can't make up its mind which tree to land in. First they fly to one, then the other, then back again. Like me, they are vagrant, having no focus. And the world is new again, and I feel clean and happy that nearly every morning for the last sixty-odd years I have popped into the world for a while and at the end of the day popped out again, and eventually the day will come when my song will end and that is all fine. I don't need to do anything, I don't need to be Sisyphus rolling his stone. I can be happy, just watching and listening and tasting the air without thinking, without doing. My beard is white, through sun or years; my head as smooth as a river stone. My autumn has come and I'm ripening – how sweet that is! How sweet a flower I'll try to be.

Coming towards me is a young man and he's staring at me, and his horns are massive and wide and proud, long horns from the kind of enormous cattle that might wander on a prairie. I'm an old, bearded white man (with goat horns at the moment) and he stares, and I start to feel that he looks challenging because he keeps staring into my eyes; he looks fierce, and I feel uncertain and I prepare myself. I wonder if he is one of the angry people who will never be satisfied until they have deleted themselves and the world around them. Then he nods and smiles at me as he passes, and I nod and smile back; and I think I'm an old man in a suit, and he's a proud and beautiful young man, and that is exact and perfect and everything is fine, and just because he is looking into my eyes, that does not mean he either wants to fuck me or kill me. So much of what I was taught as a child were lies. I do not know what is real and what isn't, and I don't care any more.

I put the flowers down on the bench next to me, take out my notebook and begin to write, knowing that in scratching these airless, scrawled, dry twisted sticks of words, once again I will struggle to capture this immensity, the light and dark of life and death. My pen a stick dragged in the sand. I wax and polish my horns. I am, after all, a Minotaur. I will gaze into things. Write about them, watch the rain.

Postscript and Acknowledgements

I need to thank my agent, the truly beautiful Robert Caskie of Robert Caskie Ltd. He grasped what I was trying to do in my work; he helped me to mould it into shape and went out to find the right people for me to work with. He is generously there whenever I need him, my first port of call. He checks in on me, usually, uncannily at exactly the right time, keeping me updated and then leaving me alone when I need to disappear down the rabbit hole into that strange isolated place where the words come from. He more than anyone enabled me to focus my life on writing. I could not have more love for him.

Elizabeth Foley, my wonderful, sensitive editor at Harvill Secker, who, seeing through my interminable repetitions and ramblings, gently brings me back by saying, 'Didn't you already say that on page 43?' and 'Tell me more about this.' A guiding hand so gentle that, as if I were a child, I can hardly see that I am being guided. Who showed such bravery in taking on my work and invested in publishing it.

Between them, these two people have changed my life and I am endlessly grateful and surprised. They liberated me, gave me wings and allowed me to have a contemplative life that no longer depends on hard physical labour.

Behind both of these people, there are teams who rarely get a mention: the fantastic Design team, led by Suzanne Dean, who do such a remarkable job, astonishing me by wrapping my words in a far better set of clothes than the overalls I imagined; the Sales team who are passionate about what they sell as they wander the country visiting booksellers bringing copies of books to share; the booksellers themselves who I get to meet at events. There is so much passion in this bookish world!

Robert Caskie's co-agents who surprise me constantly when the postman knocks on my door and hands me a cardboard box that contains copies of my book in German, or Danish, Korean or Italian or any of a dozen different languages.

And there are those I don't get to meet, the Marketing and Sales people who are constantly on the case exploring where a book fits into the market, who its readers might be, how to find them.

Every day I think myself lucky to have my partner, my wife, the gloriously talented Kate (Peggy) Hamer, who I love so much that I married her twice. We have the most wonderful creative relationship, writing our very different kinds of work at different ends of the house and meeting for coffee and dinner.

The reader may be asking, Is all this true, did it really happen that I was a tramp and a molecatcher and a gardener? The answer is yes. I hesitate to call myself a writer; I am a man who writes, certainly, but a man who writes needs to earn a living doing other things while he is creating his world. Many years ago, in the 1970s, the days before the Internet and hard disks and backups, I was unemployed for a winter, living in an isolated and unheated house on the coast of west Wales, my nearest neighbour an hour's walk away. Dyslexic, I am unable to read my own handwriting and in that winter I wrote a novel on a tiny typewriter with a worn-out ribbon that I inked myself. The manuscript, long lost along with the defunct technology, was a mess because my hands are big and the keys were childishly small. The story was set in a mythical garden where a gardener lived with various fantastic half-human creatures. Really it was about depression, loneliness and the fog that obsession creates between us. I do like hidden levels in a story. I submitted it to a literary agent who told me that she thought it beautiful but could not imagine where she could place it. Over two pages she encouraged me to keep writing, I followed her advice and wrote many things that have been left behind where they belong, so I would like to thank an agent that was never mine, who I never met, for sending me on the long journey all the way to this book.

And finally I would like to thank you, the reader, wherever you are, in whatever language you are reading this, I have nothing but praise for you who have got this far, even to the point of reading the acknowledgements! Thank you so very much for buying or borrowing this book. I hope that you enjoyed it. My very best wishes,

Marc Hamer

penguin.co.uk/vintage